A Practical Guide to Managing Clinical Trials

A Practical Guide to Managing Clinical Trials

JoAnn Pfeiffer
Cris Wells

CRC Press is an imprint of the
Taylor & Francis Group, an **informa** business

CRC Press
Taylor & Francis Group
6000 Broken Sound Parkway NW, Suite 300
Boca Raton, FL 33487-2742

First issued in paperback 2020

ISBN-13: 978-1-138-19650-6 (hbk)
ISBN-13: 978-0-367-49782-8 (pbk)

Library of Congress Cataloging-in-Publication Data

Names: Pfeiffer, JoAnn, author. | Wells, Cris, author.
Title: A practical guide to managing clinical trials / JoAnn Pfeiffer and Cris Wells.
Description: Boca Raton : CRC Press, [2017] | Includes bibliographical references and index.
Identifiers: LCCN 2016050613| ISBN 9781138196506 (hardback : alk. paper) | ISBN 9781315299792 (ebook)
Subjects: LCSH: Clinical trials--Management--Handbooks, manuals, etc. | Clinical trials--Safety measures.
Classification: LCC R853.C55 P484 2017 | DDC 610.72/4--dc23
LC record available at https://lccn.loc.gov/2016050613

Visit the Taylor & Francis Web site at
http://www.taylorandfrancis.com

and the CRC Press Web site at
http://www.crcpress.com

Contents

Preface

Over the years, as we have prepared curriculum and sought out materials for clinical research courses, we have discovered some excellent books related to clinical research and its conduct. Many have focused on the role of the investigator, and others have been directed toward research coordinators, research nurses, and the pharmaceutical or medical device industry. Still others have focused on research conducted in academic medical centers and federally sponsored programs. Yet, we struggled to find a practical book that focused on investigator/site responsibilities for the conduct of industry-sponsored clinical research—a book to meet the needs of clinical research students and to be flexible enough to meet the resource needs of clinical research professionals. As such, this book's genesis is a direct result of our frustrations in finding the "right" book for our courses.

Throughout the book, you will note that we have included "A View from India" within each of the chapters and as the focus of Chapter 10. Global clinical research is a key consideration for all types of clinical research but has special implications in regard to medical product approval across geographical and cultural boundaries. At the time of the writing of this book, India has become a powerhouse of clinical research and is a key player in the outsourcing of clinical trials from the United States. With this in mind, we called on a special colleague, Dr. Kaushal Piyush Shah, to summarize some of the salient points of difference in Indian clinical trials compared with those in the United States. We thank her for her contributions throughout the book and our clinical research programs.

Keeping in mind that this book is a snapshot in time, it is important to note that there are many changes in the pipeline for science and for regulations that will impact how clinical research will be conducted in the next years. Our hope is that you will find this book helpful in defining current practices, guidelines, and regulations, as well as defining baselines for the future.

Acknowledgments

We would like to offer a special thanks to Kaushal Piyush Shah, PhD, who has many years of experience conducting clinical research in India. In addition to authoring "A View from India" and Chapter 10, she spent countless hours reviewing the chapters and providing feedback to the authors. We truly appreciate Kaushal's dedication and generosity in this endeavor.

We could not have completed this book without our husbands. Thank you Bob Pfeiffer and Larry Wells for your incredible patience and support during this process—you made this book possible!

Authors

JoAnn Pfeiffer, DrSC, RAC, CCRA, is the director and associate clinical professor in the Clinical Research Management Program at Arizona State University. Her experience includes directing clinical trials in both academic and nonacademic environments, where she was responsible for day-to-day operations, site compliance, regulatory submissions, data and document management, clinical trial budgets, and contracts. Dr. Pfeiffer teaches master-level courses in regulatory science and clinical research management. As a subject matter expert, she presents nationally and is a published author in the areas of contract and budget management and clinical research operations.

Cris Wells, EdD, MBA, CCRP, RT(R)(M), is a clinical professor and senior director of Health-Related Programs at Arizona State University (ASU) within the College of Nursing and Health Innovation. A radiologic technologist and electrical engineer by training, Dr. Wells has a long history of overseeing health care and research projects in basic, translational, and clinical research. Prior to joining ASU, she served in various capacities in education as well as research. Her research positions included appointments as director of clinical operations in translational research, senior director of oncology clinical research, and supervisor of research administration in neurobiology. Her engineering background centered on pacemaker design when working for a medical device company. She also served as the consumer representative on the Food and Drug Administration Medical Device Advisory Panel for 4 years. Professor Wells' current interests are focused on workforce development and curriculum design of emerging research and health care professions, including biorepository administration and health care compliance.

List of Acronyms

ADR	adverse drug reaction
AEs	adverse events
ALCOA	attributable, legible, contemporaneous, original, accurate
BIMO	bioresearch monitoring
BLA	biologics license application
CAPA	corrective action/preventive action
CBC	complete blood count
CBER	Center for Biologics Evaluation and Research
CDA	confidential disclosure agreement
CDER	Center for Drug Evaluation and Research
CDRH	Center for Devices and Radiological Health
CDSCO	Central Drug Standard Control Organization
CFR	Code of Federal Regulations
cGMP	current good manufacturing practices
cGTP	current good tissue practices
CIOMS	Council for International Organizations of Medical Sciences
CMS	Centers for Medicare and Medicaid Services
CoC	Certificate of Confidentiality
CPT	current procedure terminology
CRA	clinical research associate
CRC	clinical research coordinator
CQMP	clinical quality management plan
CRFs	case report forms
CRO	clinical research organization
CTRI	Clinical Trials Registry–India
CV	curriculum vitae
D & C Act	Drug & Cosmetics Act 1940 & Drug and Cosmetic Rules 1945 (India)
DBT	Department of Biotechnology
DCC	Drug Consultative Committee
DCGI	Drug Controller General of India
DMAC	Drugs Medical Advisory Committee
DOJ	Department of Justice
DSMB	Data Safety Monitoring Board
DTAB	Drugs Technical Advisory Board
EC	ethics committee
eCRF	electronic case report form
EDC	electronic data capture system
EHR	electronic health record
EKG	electrocardiogram
EMA	European Medicines Agency
ePHI	electronic protected health information

ESCs	embryonic stem cells
EU	European Union
FDA	U.S. Food and Drug Administration
FIH	first in human
Form FDA 1572	Statement of the Investigator
Form FDA 482	Notice of Inspection
Form FDA 483	Statement of Observations
FWA	Federal Wide Assurance
GCP	good clinical practice
HCT/P	human cell tissue/products
hESC	human embryonic stem cells
HHS	U.S. Department of Health and Human Services
HIPAA	Health Insurance Portability and Accountability Act
HIV	human immunodeficiency virus
IB	investigator brochure
ICD	informed consent document
ICFs	informed consent forms
ICH	International Conference of Harmonisation
ICMR	Indian Council of Medical Research
IDE	Investigational Device Exemption
IEC	independent ethics committee
IND	Investigational New Drug
IP	intellectual property
iPSC	induced pluripotent stem cells
IRB	institutional review board
IV	intravenous
LAR	legally authorized representative
LCD	local coverage determination
MCI	Medical Council of India
MDUFA	Medical Device User Fee and Modernization Act
MOHFW	Ministry of Health & Family Welfare
MRI	magnetic resonance imaging
MTP	medical termination of pregnancy
NABH	National Accreditation Board for Hospitals and Health Care Providers
NAI	no action indicated
NCD	national coverage determination
NCI	National Cancer Institute
NDA	New Drug Application
NGSCR	National Guidelines for Stem Cell Research
NIDPOE	Notice of Initiation of Disqualification Proceedings and Opportunity to Explain
NIH	National Institutes of Health
NSR	nonsignificant risk
OAI	official action indicated
OCP	Office of Combination Products

OCR	Office for Civil Rights
OHRP	Office of Human Research Protection
OIG	Office of the Inspector General
PHI	protected health information
PHS	public health service
PI	principal investigator
PIS	patient information sheets
PMA	premarket approval
PMOA	primary mode of action
QA	quality assurance
QC	quality control
QCI	Quality Control of India
RA	rheumatoid arthritis
RBM	risk-based monitoring
RCA	root cause analysis
SADR	serious adverse drug reaction
SAEs	serious adverse events
SE	schedule of events
SOC	standard of care
SOP	standard operating procedure
SR	significant risk
SSC	somatic stem cells
URL	Uniform Resource Locater
VAI	voluntary action indicated
WHO	World Health Organization

1 Rules, Roles, and Responsibilities

Clinical research is a complex, multidisciplinary industry that integrates science, business, law, ethics, and health care. It is regulated by multiple government agencies and is conducted for a variety of purposes. As such, the roles of individuals within the industry are diverse and multifaceted. However, overall, the clinical research environment is divided into four areas that have distinct but overlapping responsibilities and goals. These areas or roles include regulatory agencies, sponsors, investigators, and ethical/institutional review boards (IRBs). Each encompasses a scope of responsibilities intertwined with regulations, ethics, best practices, and professional standards.

This book focuses on the operational conduct of clinical trials at the research site for the purpose of collecting data on the safety and efficacy of new investigational products, which will, in turn, be submitted to the Food and Drug Administration (FDA) by a sponsor for marketing approval. These clinical trials are often referred to as industry-sponsored studies because pharmaceutical or device companies "sponsor" research to prove the safety and effectiveness of their products in anticipation of marketing the product to the public. The purpose of this chapter is to review the rules, standards, regulations, and best practices that are part of good clinical practices (GCPs), then to explore the roles and responsibilities of sponsors, investigators, and ethical review boards (or IRBs) as they relate to the conduct and operations of industry-sponsored clinical trials.

GCP, REGULATIONS, AND OVERSIGHT

Good clinical practice, or GCP, is a combination of standards, regulations, guidance documents, and best practices that, when combined, ensure the ethical and scientific conduct of clinical trials. In this book, we focus on FDA regulations and guidance documents, Office for Human Research Protections (OHRP) regulations and guidance documents, and International Conference on Harmonisation (ICH) guidelines, as we discuss GCP related to conducting clinical research.

Food and Drug Administration

Although the U.S. FDA's regulatory functions began with the passage of the 1906 Pure Food and Drugs Act, its current name and regulatory functions were officially established under the Federal Food, Drug, and Cosmetic Act of 1938. The Act was passed as a direct reaction to the Elixir Sulfonamide Tragedy in which more than 100 children and adults ingested a sweet-tasting elixir that had been sold as a drug but which also contained antifreeze. Their deaths spurred legislation toward the oversight of various drugs. Besides bringing cosmetics and medical devices under its control, the Act required premarket approval of all new drugs, mandating that a manufacturer would have to prove to the FDA that a drug was safe before it could be sold. Section 505(i) of the Act gives the FDA authority for oversight of clinical investigations to test for safety and effectiveness of investigational products.

Today, the FDA regulates food, drugs, biologics, medical devices, electronic products that emit radiation, cosmetics, veterinary products, and tobacco. Of specific relevance to this book is the FDA oversight of drugs, biologics, and medical devices in issuing regulations that permit and govern the investigational use of drugs and devices to determine their safety and effectiveness before approval or clearance. Title 21 of the Code of Federal Regulations focuses on the FDA regulations relevant to market approval. These regulations include the following:

- 21 C.F.R. Part 312: Investigational New Drug Application
- 21 C.F.R. Part 812: Investigational Device Exemptions
- 21 C.F.R. Part 50: Protection of Human Subjects
- 21 C.F.R. Part 11: Electronic Records, Electronic Signatures
- 21 C.F.R. Part 54: Financial Disclosure by Clinical Investigators
- 21 C.F.R. Part 56: Institutional Review Boards

As part of its approval of medical products, the FDA regularly performs inspections of research sites to ensure compliance to GCP. At these inspections, the FDA inspector reviews and discusses his/her findings with the investigator and the research team. If deficiencies are found, they are summarized on a form entitled "Inspectional Observations," otherwise known as a *Form 483* (Appendix A). Inspections are conducted through the FDA's Bioresearch Monitoring Program (BIMO) for the following reasons:

- To verify the accuracy and reliability of data that have been submitted to the agency;
- As a result of a complaint to the agency about the conduct of the clinical trial at a particular research site;
- In response to sponsor concerns;
- Upon termination of the research site;
- During ongoing clinical trials to provide real-time assessment of the investigator's conduct of the trial and protection of human subjects;
- At the request of an FDA review division; and
- Related to certain classes of investigational products that the FDA has identified as products of special interest in its current work plan (i.e., targeted inspections based on current public health concerns) (U.S. Department of Health and Human Services [HHS], FDA, 2010a, p. 3).

Inspections, audits, monitoring visits, and corrective action plans will be explored further throughout the book but will be emphasized in Chapter 6.

Office of Human Research Protections

The OHRP is part of the Office of the Assistant Secretary for Health within the U.S. Department of HHS. It is tasked with overseeing compliance related to the rights of human subjects of research conducted or supported by HHS under Section 289 of the Public Health Service Act. Specifically, OHRP oversees the regulations codified in 45 C.F.R. Part 46, Protection of Human Subjects, informally known as the "Common Rule," and research that is conducted or supported by various government agencies or departments that have adopted the Common Rule.

The Common Rule applies to all research involving human subjects that is conducted or supported by HHS (grants or other funding mechanisms) or is conducted in an institution that has agreed to assume responsibility for the research in accordance with HHS regulations, regardless of the source of funding. It was published in 1991 and adopted by 15 Federal departments and agencies. These agencies have adopted section numbers and language that are identical to those found in the HHS regulations, codified as 45 C.F.R. Part 46, subpart A (U.S. HHS, n.d.). This section outlines the basic provisions for IRBs, informed consent, and Assurances of Compliance. An example of a clinical trial that OHRP would oversee might include a human subject clinical trial comparing the nutritional benefit of carrots versus broccoli, sponsored by funding through the Department of Agriculture (as a department that has adopted the Common Rule).

The Common Rule is based on the "Ethical Principles and Guidelines for the Protection of Human Subjects of Research," otherwise known as the Belmont Report (Appendix A). The ethical principles of respect for persons, beneficence, and justice found in the Belmont Report underpin the concepts of informed consent, subject benefit and risk, and fair distribution found within the regulations. The Common Rule is found in the following regulatory citation:

- 45 C.F.R. Part 46: Protection of Human Subjects

Although the regulations found in 45 C.F.R. Part 46 are very similar to those found in 21 C.F.R. Part 50, the OHRP and the FDA each has different scopes and oversight. The FDA's scope is aligned with the purpose of medical product approval for marketing, and the OHRP oversees research conducted or sponsored by government agencies or departments that have adopted the Common Rule. However, there may be times when a clinical trial may fall under both 45 C.F.R. Part 46 and 21 C.F.R. Part 50. As an example, the National Cancer Institute may sponsor a clinical trial in collaboration with a pharmaceutical company that is in the midst of conducting studies under an Investigational New Drug (IND) application for the drug. In this case, the research site is obligated to follow both sets of regulations.

Similar to FDA inspections, the OHRP conducts inspections through its Division of Compliance Oversight. The OHRP will conduct a for-cause inspection in response to receipt of "substantive written allegations or indications of noncompliance with HHS regulations" from research subjects, institutional officials, or others who may have concerns in regard to compliance with Title 45 regulations under the OHRP jurisdiction (U.S. OHRP, 2009).

International Conference on Harmonisation

The ICH evolved from a meeting held in 1990 to explore the possibility of "harmonizing" pharmaceutical regulations throughout Europe, Japan, and the United States. The objective of the resulting E6 Guideline for Good Clinical Practice is to provide a "unified standard for the European Union (EU), Japan and the United Sates to facilitate the mutual acceptance of clinical data by the regulatory authorities in these jurisdictions" (ICH, 1996, p. 1).

According to the ICH (1996), "good clinical practice (GCP) is an international ethical and scientific quality standard for designing, conducting, recording, and reporting trials that involve the participation of human subjects" (p. 1). Although not a regulation, the ICH Guidance has been adopted as a standard within GCP by the FDA and by other countries throughout the world. This is particularly important because of the global nature of clinical trials. Many studies include research sites within and outside U.S. geographic boundaries, and sponsors expect that research sites follow a common set of rules and guidance. (The E6 Document can be found in its entirety within Appendix A.)

Additional Laws and Guidance

Because clinical research is conducted in a variety of venues and as a business entity, there are a multitude of other regulations (national and state), industry standards, and best practices to be considered when engaged in clinical research. As we continue through this book, we will be discussing these other rules along with the previously mentioned regulations and guidelines as they impact the conduct of clinical research.

SPONSORS

A sponsor is normally the developer of a drug, biologic, or medical device and oversees the investigational product's growth from initial identification of the potential product through manufacturing and testing of the product in people. It is the sponsor that requests permission from the FDA to conduct research with its drug, device, or biologic for eventual marketing.

More formally, the ICH (1996) defines *sponsor* as "an individual, company, institution, or organization which takes responsibility for the initiation, management, and/or financing of a clinical trial" (p. 7). Alternatively, the FDA defines *sponsor* as "a person who initiates a clinical trial, but who does not actually conduct the trial, i.e., the test article is administered or dispensed to or used involving, a subject under the immediate direction of another individual. A person other than an individual (for example, corporation or agency) that uses one or more of its own employees to conduct a clinical trial it has initiated is considered to be a sponsor (not a sponsor-investigator), and the employees are considered to be investigators" (FDA IND Definitions and Indications, 2015). In this book, we will focus on the description or definition of *sponsor* as a company, institution, or an organization, rather than as an individual.

Sponsors have four main areas of responsibilities in the development of a new medical product, including preclinical research, manufacturing, clinical trials, and postapproval obligations. Overall, in the case of a drug approval, the sponsor is responsible for submission of an IND application to the FDA, the conduct of clinical trials, and submission of a New Drug Application (NDA) for drugs or the Biological License Application for Biologics (BLA). Alternatively, for the clinical evaluation of a device that has not been cleared for marketing, the sponsor must have an approved Investigational Device Excemption (IDE) prior to clinical trial initiation. For the purposes of this chapter and this book, we are interested in the responsibilities and obligations of the sponsor and clinical research (rather than preclinical research, manufacturing, or postapproval obligations).

According to 21 C.F.R. 312.50 (IND), sponsors assume the following responsibilities during the conduct of clinical:

- Selecting qualified investigators;
- Providing investigators with the information they need to conduct an investigation properly;
- Ensuring proper monitoring of the investigation(s);
- Ensuring that the investigation(s) is conducted in accordance with the general investigational plan and protocols contained in the IND;
- Maintaining an effective IND with respect to the investigations; and
- Ensuring that the FDA and all participating investigators are promptly informed of significant new adverse effects or risks with respect to the investigational drug or product (FDA IND General Responsibilities of Sponsors, 2015).

For investigations relating to an IDE (21 C.F.R. 812.40), sponsors are responsible for selecting qualified investigators, providing investigators with the information needed to conduct the investigation properly, and ensuring proper monitoring of the investigation similarly to the IND regulations above. Additionally, sponsors assume the following responsibilities for an IDE:

- Ensuring IRB review/approval;
- Submitting an IDE application to the FDA;
- Ensuring that any reviewing IRB and the FDA are informed about any significant new information about an investigational device (FDA IDE General Responsibilities of Sponsors, 2015).

In summary, sponsors are responsible for choosing investigators who are qualified by training and education; ensuring that investigators have all the information needed to conduct the clinical trial; monitoring the progress of the clinical trial for safety, compliance, record keeping, and investigational product handling; and reporting to the FDA.

Contract Research Organizations

A sponsor may transfer all or "certain" IND/IDE obligations to another organization known as a contract research organization (CRO). According to 21 C.F.R. 312.3 (b), a CRO "means a person that assumes, as an independent contractor with the sponsor, one or more of the obligations of a sponsor, for example, design of a protocol, selection or monitoring of an investigation, evaluation of reports, and preparation of materials to be submitted to the Food and drug Administration" (2015). A CRO is typically not a "single" person, but rather a specialized company or entity that contracts specifically with a sponsor to perform certain obligations.

In keeping with 21 C.F.R. 312.50 (2015), any transfer of obligations to a CRO must be in writing. If not all the obligations are transferred, the written document must describe each of the obligations that is being assumed by the CRO. If any obligation is not described in the document as being transferred, then the responsibility for the obligation remains with the sponsor. Once an obligation is delegated, the CRO also assumes the responsibility and consequences of compliance and noncompliance of the obligations.

Because there may be confusion as to whether the sponsor or the CRO is responsible for handling various aspects of a clinical trial, it is important for an investigator and his/her research team to understand which obligations have been transferred to a CRO and which obligations stay with the sponsor. For example, a CRO may be responsible for contracting with the research sites, but the sponsor may be responsible for monitoring the research sites. Unfortunately, confusion in regard to responsibilities can cause duplication of effort, delays in site payments, compliance issues, safety concerns, and other consequences. Throughout this book, the authors will refer to "sponsors" with the assumption that sponsor and CRO may be interchangeable.

Sponsor Personnel

The sponsor hires employees who will support the proper operational conduct of its research studies. Although most of these positions are "behind the scenes," there is one position that is integral to the overall research team and interacts extensively with the investigator and the research site—the monitor.

The monitor is also known as a clinical research associate (CRA). He or she is the sponsor's representative in visiting and working with research sites to ensure successful operation of the clinical trial. Compliant with 21 C.F.R. 312.53 (FDA IND Responsibilities of Sponsors and Investigators, 2015), "a sponsor shall select a monitor qualified by training and experience to monitor the progress of the investigation." Some of the clinical operations and monitoring activities within the CRA position include clinical trial planning and start-up, clinical trial development and contracting, clinical trial conduct oversight, clinical trial closeout, postclinical trial activities, training, and project management. The monitor has a major influence on the conduct and success of a clinical trial.

The sponsor's research team may also include medical research associates, medical monitors (physicians), biostatisticians, medical writers, regulatory affairs specialists, quality assurance managers, data entry personnel, and others.

INVESTIGATORS

An investigator is a person qualified by experience and education to lead a clinical trial at a research site. In accordance with 21 C.F.R. 312.3, "Investigator means an individual who actually conducts a clinical investigation (i.e., under whose immediate direction the drug is administered or dispensed to a subject). In the event an investigation is conducted by a team of individuals, the investigator is the responsible leader of the team" (FDA IND Definitions and Indications, 2015).

An investigator must be familiar with the investigational product, whether it is a drug, device, or biologic, and be qualified by training and education as an appropriate expert in a therapeutic area to conduct a clinical trial. As an example, an investigator who has been trained as an internist would most likely not be qualified to act as an investigator for a clinical trial involving dental implants because his or her training and expertise is not specific to dentistry. In most cases, the investigator who leads the research team is called the principal investigator, and other investigators on the research team may be termed subinvestigators. Although it is not required that the investigator hold a medical license (or be a physician), "a qualified physician or dentist should be listed as a sub-investigator for the trial and should be responsible for all trial-related medical or dental decisions" (U.S. HHS FDA, 2010b, p. 5).

An investigator is responsible for the following:

- Ensuring that an IRB or ethics review board approves the protocol and informed consent document before the trial begins;
- Complying with the IRB-approved protocol;
- Reading and understanding the information provided by the sponsor (investigator brochure, preclinical data, protocol, informed consent form, and other sponsor-provided information and training);
- Obtaining informed consent from subjects prior to any trial-related procedures;
- Maintaining investigational product accountability;
- Supervising the conduct of the trial;
- Maintaining adequate and accurate subject and trial records and making them available for inspection(s);
- Reporting adverse experiences to the IRB according to the protocol and regulations.

The investigator commits to these responsibilities and all other requirements regarding the obligations of investigators found in Part 9 of the "Statement of the Investigator" (otherwise known as the Form FDA1572) and all other pertinent requirements in 21 C.F.R. Part 312.

The FDA may disqualify an investigator if she or he "has repeatedly or deliberately failed to comply with applicable regulatory requirements or has repeatedly or deliberately submitted false information to the sponsor or to the FDA" (U.S. FDA, 2011, para. 2). The consequences of an investigator being disqualified include not being eligible to receive any investigational drugs, biologics, or devices and not being eligible to conduct any clinical trials in supporting an application for research or marketing of medical products regulated by the FDA. Once on the disqualification list, which is published on the FDA website, an investigator's name stays there forever, even if she or he is reinstated and is allowed to do research again. If an investigator has been reinstated, his or her status is noted next to his or her name.

The investigator is ultimately responsible for the conduct of a clinical trial at the research site and the integrity, health, and welfare of the human subjects during the clinical trial. As such, the investigator must understand the consequences of signing the Form FDA 1572 or agreeing to the "Statement of the Investigator."

Research Site Personnel

Although the investigator assumes full responsibility for the conduct of a clinical trial, she or he may not be able to complete all the tasks required for success of the clinical trial without assistance from other individuals trained in conducting clinical trials and/or other specialized operations. The investigator may delegate tasks to individuals who have the appropriate educational background, licensure, and/or training. For example, a clinical trial may require a nuclear imaging scan that requires specialized knowledge of a particular isotope. The investigator would delegate the task to licensed radiologists and nuclear medicine technologists, who have been trained to conduct the scan in compliance with regulations and the protocol. Although the investigator delegates these tasks, she or he is still responsible for ensuring the tasks are done according to the protocol, GCP, and other rules and regulations.

Often times, investigators hire clinical trial/research coordinators to assist them with tasks related to subject screening, clinical trial maintenance, and clinical trial closure. Additionally, the investigator may enlist a pharmacist to assist with investigational product accountability, a data coordinator to abstract data for submission of case report forms, and an IRB coordinator to manage the submissions and maintenance of IRB applications, continuations, and closures. However, at the risk of repetition, the investigator is ultimately responsible for the overall conduct of the clinical trial at the research site.

ETHICAL (INSTITUTIONAL) REVIEW BOARDS

An ethical review board or IRB is a committee formed by an institution or an independent corporate entity to oversee the approval of research. Its goal is to protect human subjects in research. The FDA regulations (21 C.F.R. 56.102(g)) define an IRB as "any board, committee or other group formally designated by an institution to review, approve the initiation of and conduct periodic review of biomedical research involving human subjects. The primary purpose of such review is to assure the protection of the rights and welfare of the human subjects" (FDA Institutional Review Boards Definitions, 2015). The regulations also require IRBs to follow written procedures and to document their decisions and retain the documentation appropriately. Institutional review boards are addressed through the following sets of regulations:

- 21 C.F.R. Part 50: Protection of Human Subjects (FDA)
- 21 C.F.R. Part 56: Institutional Review Boards (FDA)
- 45 C.F.R. Part 46: Protection of Human Subjects (Common Rule)

According to the Code of Federal Regulations, an IRB must have at least five members selected to represent an appropriate diverse exemplification of profession, ethnic background, gender, community knowledge, and scientific background. It is the IRB's responsibility to determine whether a clinical trial should be done and what manner of informed consent is adequate and appropriate.

To determine whether a clinical trial should be conducted, the IRB must consider different aspects of the clinical trial: its scientific validity, the balance between the risks to subjects and the benefits to subjects, and the subject selection process (Belmont Principles). Although the IRB is not typically assigned to determine the scientific merit of a clinical trial, it is responsible for making sure that the scientific methods used in the clinical trial are valid. The IRB also reviews the informed consent process (protocol and IRB application), subject recruitment materials, the informed consent document, and the privacy and confidentiality aspects of the process.

In summary, the IRB is responsible for protecting the rights and welfare of human subjects in clinical trials by assuming the following responsibilities:

- Ensuring that risks to subjects are minimized and reasonable in relation to anticipated benefits;
- Ensuring that subject selection is equitable;
- Reviewing the protocol, informed consent document, investigator brochure, recruitment and advertising materials;
- Providing initial and continuing review of ongoing clinical trials at various intervals (including review of adverse events and other trial-related events);
- Approving, disapproving, and/or terminating a trial.

IRBs and Personnel

In many institutions (hospitals, medical centers, educational or academic sites, or others), the IRB is located in a separate department within the Research Administration Unit. The department is typically staffed with an IRB administrator, IRB coordinator(s), and/or administrative assistant(s) to assist with the administrative record keeping, standard operating procedures, and other regulatory and organizational requirements of the actual Board. The members of the Board are usually volunteers from the organization, community, and/or other members who meet the qualifications specified in the regulations. The time commitment for members of these Boards is considerable and may be compensated by the organization by reallocation of work duties or by a stipend.

Internal IRBs may not be practical for some organizations that do not have the resources to host their own IRB. In this case, there are commercial IRBs that may meet an organization's needs. These IRBs are held to the same regulatory and ethical standards as internal IRBs but may be owned as a commercial entity. Commercial IRBs may offer the advantage of resources, convenience, expertise, and/or consistency across a multiple-site clinical trial. Commercial IRBs are business entities and provide services in exchange for monetary compensation. Their employees and consultants are compensated per business standards.

A VIEW FROM INDIA

REGULATORY OVERSIGHT

The Central Drugs Standard Control Organization (CDSCO) within the Ministry of Health & Family Welfare (MOHFW) is the national regulatory body for drugs, cosmetics, and medical devices in India. The CDSCO is equivalent to the FDA of the United States. The Drugs Controller General of India (DCGI) is the head of the CDSCO and is equivalent to the FDA Commissioner. The DCGI approves new drugs, cosmetics and medical devices, and clinical trials. It also monitors the quality and efficacy of pharmaceutical products available in the market.

THE DRUGS AND COSMETICS ACT 1940 AND THE DRUGS AND COSMETICS RULES 1945 (D & C ACT)

The Drugs and Cosmetics Act 1940 and the Drugs and Cosmetics Rules 1945 (D & C Act) are part of India's national legislation that regulates the import, manufacture, distribution, and sale of drugs, cosmetics, and medical devices in India. These regulations are amended from time to time. Schedule Y, the regulation for clinical research was introduced in the D & C Act in 1988. The rules that apply for the conduct of Clinical Trials are listed in Appendix A.

SCHEDULE Y

Schedule Y is equivalent to the IND regulations described in 21 C.F.R. Part 312. Clinical trials in India are conducted under the rules and regulations contained in Part X-A: "Import or Manufacture of new drug for clinical trials or marketing," of the D & C Act. The recent 2013 version of Schedule Y has three major sections and 12 appendices. These are listed in Appendix A.

ETHICAL AND REGULATORY GUIDELINES

The Indian Council of Medical Research (ICMR) issued the Ethical Guidelines for Biomedical Research on Human Subjects in 2000, which were revised in 2006. Based on ethical codes, such as the Nuremburg Code, and international declarations, such as the Declaration of Helsinki, the ICMR guidelines are recognized by the OHRP. The CDSCO released Indian GCP guidelines in 2000, and they have been endorsed by the Drug Technical Advisory Board (highest technical body under the D & C Act) for implementation in conducting all biomedical research in India.

To decrease the time to approval of global clinical trials, the DCGI created two categories of applications:

- Category A: Clinical trial applications for clinical trials approved in countries with competent, mature regulatory systems, such as the United States, Germany, United Kingdom, Switzerland, Australia, Japan, South

Africa, Europe, and Canada are fast-tracked for a goal of approval within 2–4 weeks.
- Category B: Approvals for all other clinical trial applications undergo a routine approval process with a goal of approval within approximately 12 weeks.

The Clinical Trials Registry (CTRI)—India

All clinical trials being conducted in India must be registered in the Clinical Trials Registry–India (CTRI) found at the following website: www.cdsco.nic.in. The CTRI is equivalent to clinicaltrials.gov run by the U.S. National Library of Medicine. In support of the registry, editors of 11 major Indian biomedical journals declared that only registered trials would be considered for publication in the journals (ICMR, n.d.).

Sponsor and Sponsor Personnel

All the roles and responsibilities of the sponsor and the sponsor personnel mentioned previously in this chapter are similar to those in India.

Investigators and Research Site Personnel

The roles and responsibilities of the investigator and the research site personnel mentioned previously in the chapter are similar to those in India. The investigator signs the Investigator Undertaking, which is similar in format and equivalent in purpose to the U.S. Form FDA 1572. The Investigator Undertaking is specified in Appendix VII of Schedule Y.

Ethical (Institutional) Review Boards, Personnel, and Accreditation

The goals and responsibilities of IRB and Independent Ethics Committee (IEC), as mentioned previously in the chapter, are similar to those in India. The IRBs and IECs are addressed through the regulations mentioned in Appendix VIII of Schedule Y. According to Appendix VIII of Schedule Y, an IRB/IEC must have at least seven members with representation from basic medical scientists, clinicians, legal experts, social scientist/representation of nongovernmental organization/philosopher/ethicist/theologian, and community lay person. The recent (2013) amendment to the D & C Act introduced the Rule 122 DD: Registration of Ethics Committee, whereby all ethics committees must be registered with the licensing authority (DCGI) (Government of India, MOHFW, 2013).

CHAPTER REVIEW

In this chapter, we reviewed GCPs, the regulations, and the respective agencies or organization responsible for clinical trials. We then explored the roles and responsibilities of sponsors, investigators, and ethical review boards (or IRBs). Finally, we reviewed some of the different aspects of regulations within India.

APPLY YOUR KNOWLEDGE

1. PharmaXYZ recently contacted the investigator of a multi-investigator dermatology practice/research site to determine her interest in a diabetes clinical trial. From the perspectives of the sponsor, investigator (research site), and the IRB, what questions should be asked about the investigator's participation in the diabetes clinical trial, and why?
2. Investigator ABC has a grant from the National Cancer Institute to conduct a clinical trial for an investigational drug that has an IND through the FDA. What set or sets of federal regulations should the research site follow, and why?

REFERENCES

FDA IDE General Responsibilities of Sponsors, 21 C.F.R. § 812.40 (2015).

FDA IND Definitions and Indications, 21 C.F.R. § 312.3 (2015).

FDA IND General Responsibilities of Sponsors, 21 C.F.R. § 312.50 (2015).

FDA IND Responsibilities of Sponsors and Investigators, 21 C.F.R. § 312.53 (2015).

FDA Institutional Review Boards Definitions, 21 C.F.R. § 56.102 (2015).

Government of India, Ministry of Health & Family Welfare. (2013). *Notification*. Retrieved from http://cdsco.nic.in/writereaddata/G.S.R%2072(E)%20dated%2008.02.2013.pdf

Indian Council of Medical Research. (n.d.). Clinical trials registry—India. Retrieved from http://ctri.nic.in/Clinicaltrials/cont1.php

International Conference on Harmonisation. (1996). *Guideline for good clinical practice E6 (R1)*. Retrieved from International Conference on Harmonisation website: http://www.fda.gov/downloads/Drugs/GuidanceComplianceRegulatoryInformation/Guidances/ucm073122.pdf

U.S. Department of Health and Human Services. (n.d.). Federal policy for the protection of human subjects ('Common Rule'). Retrieved from http://www.hhs.gov/ohrp/humansubjects/commonrule/

U.S. Department of Health and Human Services, Food and Drug Administration. (2010a). *Information sheet guidance for IRBs, clinical investigators, and sponsors: FDA inspections of clinical investigators*. Retrieved from http://www.fda.gov/downloads/RegulatoryInformation/Guidances/UCM126553.pdf

U.S. Department of Health and Human Services Food and Drug Administration. (2010b). *Information sheet guidance for sponsors, clinical investigators, and IRBs: Frequently asked questions—Statement of investigator (Form FDA 1572)* [Guidance document]. Retrieved from http://www.fda.gov/downloads/RegulatoryInformation/Guidances/UCM214282.pdf

U.S. Food and Drug Administration. (2011). Disqualified/totally restricted list for clinical investigators. Retrieved from http://www.fda.gov/ICECI/EnforcementActions/DisqualifiedRestrictedAssuranceList/ucm131681.htm

U.S. Office for Human Research Protections. (2009). Compliance oversight procedures for evaluating institutions. Retrieved from http://www.hhs.gov/ohrp/compliance/evaluation/index.html

2 Products, Protocols, and Pretrial Preparation

Pretrial preparation sets the stage for managing the many complex factors that arise during clinical trials. Additionally, the process of pretrial preparation offers research sites the opportunity for process consistency, economies of scale, and technology integration across multiple studies. In this chapter and throughout the book, we will be referring to standard operating procedures (SOPs), checklists, forms, and other tools that offer research sites opportunities for efficiency.

In this chapter, we initiate our discussion by reviewing drug development and medical device development processes as they relate to the Food and Drug Administration (FDA) approval process and clinical research conducted at the site. Then, we explore sponsor research site selection, confidential disclosure agreements (CDAs), clinical trial feasibility, institutional review board (IRB) applications/submissions, and essential documents.

DRUG AND MEDICAL DEVICE DEVELOPMENT

Although drugs and medical devices have many commonalities in their paths to approval, the requirements for clinical trials to support the approval process vary significantly with medical devices. Not all medical devices need to undergo controlled clinical trials to gain regulatory approval. In the following pages, we will review the regulatory paths for drugs and medical devices while focusing on the clinical trial aspect of the process.

INVESTIGATIONAL NEW DRUG APPLICATIONS

When the sponsor of an investigational drug or biologic has gathered enough data through preclinical testing to determine that the investigational product exhibits pharmacological activity that warrants commercial development and that it is reasonably safe for initial use in humans, the sponsor must submit an application to the FDA to request permission to begin clinical trials. This application is referred to as an Investigational New Drug (IND) application, which is technically an application to the FDA to request an exemption from current Federal law that requires that a drug be part of an approved marketing application before it can be transported across state boundaries. The exemption allows the sponsor to ship investigational drug to clinical investigators in multiple states.

Once the sponsor submits an IND application, the FDA has 30 days to review the application for safety concerns before the sponsor can begin clinical trials. If the sponsor has not received communication from the FDA within the 30 days, the sponsor may begin clinical trials as it proposed in the application. However, most sponsors confirm that they can move forward by contacting the FDA if they have not received notice within the 30-day period.

Clinical trials conducted under an IND are typically categorized into phases, which implies that they are conducted consecutively. However, in practice, these phases are very fluid, they can overlap, and trials in one phase can be conducted with trials in other phases. As an example, it is not unusual for a clinical trial to be classified as a phase 1/2, study where the purpose is to determine efficacy and toxicity from a late phase 1 study into an early phase 2 study. The common phases and their descriptions follow:

- Phase 0: Otherwise known as an Exploratory IND, phase 0 studies offer sponsors a means to explore more efficient development of investigational drugs by allowing testing of small amounts of an investigational drug over a very short period (less than 7 days) in human subjects. Unlike the other phases, there is no therapeutic or diagnostic intent for this type of study.
- Phase 1: The purpose of phase 1 studies is to collect information on the safety and appropriate dosage for an investigational drug. Phase 1 researchers examine how the investigational drug works in the body, evaluate the side effects associated with dosage escalation, and start to collect information about effectiveness. Typically, these types of studies are short in duration (several months in length) and include 20 to 80 healthy volunteers as subjects, unless the investigational drug is intended for the use of cancer patients. In the cancer patient population, phase 1 studies are conducted within the group of patients suffering from the type of disease, such as liver cancer or glioblastoma. Phase 1a studies are usually single ascending dose studies, whereas phase 1b studies are usually multiple ascending dose studies, but there is variability in these descriptions across research sites and investigators.
- Phase 2: Phase 2 studies include several hundred human research subjects who are suffering from the disease or condition for which the investigational drug is being developed, and the studies have a duration of up to 2 years. These studies provide information on the safety and side effects of the investigational drug but, are not large enough to be statistically significant in regard to whether the investigational drug will be effective or beneficial to the larger intended therapeutic audience.
- Phase 3: Often known as pivotal studies, phase 3 studies are designed to demonstrate whether or not an investigational product offers a benefit to a specific population that has a disease or condition. These studies often run 1 to 4 years and include hundreds to thousands of participants. Because of the larger clinical trial population, phase 3 studies provide large amounts of data on safety and efficacy, specifically related to the population being studied.
- Phase 4: Otherwise known as postmarketing studies, these studies take place after an investigational drug has been approved and marketed. These studies are generally conducted to monitor long-term safety and efficacy under "real-life" situations and include thousands of participants.

At the research site, each phase of a clinical trial has unique characteristics that should be considered during pretrial preparation. Phase 0 trials are very specialized and typically require oversight by a clinical investigator as well as a translational scientist (can be one and the same). Additionally, the research site needs access to laboratory and other medical imaging or testing facilities and 24-hour care. Similarly, phase 1 studies are short in duration, are very intense, and require direct oversight by an investigator. Phase 1 studies require frequent tests and monitoring, which in turn require a high ratio of staff (research coordinators and medical professionals) to research subject. Phase 1 research sites generally have specialty (calibrated) medical equipment, temperature- and humidity-controlled drug storage, on-site blood chemistry processing, laboratories, infusion chairs, and inpatient-style hospital beds. Although fewer subjects are required for these types of studies, it may be difficult to recruit "healthy" research subjects, depending on the definition of "healthy" and on the demographic of the population being recruited.

Unlike phase 1 studies, phase 2 studies do not typically require a dedicated unit that is monitored full time, but they are usually conducted within a hospital or outpatient setting where subjects are (already) being treated for their condition or illness. These studies require safety and effectiveness oversight, but the staff/clinical trial subject ratio is not as high as in phase 1 studies. Subjects are recruited through physician referral, through patient advocacy or support groups, and through foundations such as the Susan G. Komen Foundation.

Phase 3 studies are often conducted in outpatient settings and include larger and more diverse populations than phase 2 studies. Because phase 3 studies are randomized and blinded (meaning that subjects are assigned to different arms of a study by chance and do not know whether they are assigned to an investigational product), they require a dedicated clinical trial coordinator who works with the research pharmacist to ensure accurate distribution and documentation of the investigational product. Additionally, recruitment and screening of subjects can be particularly onerous, requiring physician referral networks, dedicated recruitment specialists, and advertising funds.

Phase 4 studies are postmarketing studies. It is important to remember that they are conducted after the FDA has approved a drug or biologic to be sold for an indicated treatment in a specified population. This means that the New Drug Application has been approved by the Center for Drug Evaluation and Research for investigational drugs or that a biologic has been approved through the Biologics License Application by the Center for Biologics Evaluation and Research. Because these types of studies have a broader array of clinical trial designs and have different goals from premarketing studies, there may be confusion on the actual purposes of a phase 4 clinical trial. Depending on the investigator's interest and preference, these studies may not be desirable for some research sites.

Investigational Device Exemptions

Medical devices are divided into three categories based on the level of risk to users. They are then grouped according to the medical area of use. Class I devices offer minimal risk to a user and fall under General Controls, the baseline requirements that apply to all medical devices. An example of a Class I device is a tongue depressor. Class II and Class III devices are more complex devices and include moderate and high risk, respectively.

Class II devices, such as infusion pumps, must follow Special Controls and labeling, mandatory performance standards, and postmarketing surveillance. Class III devices support or sustain human lives and have the strictest guidelines because they pose the greatest risk. Besides following the Class I and Class II guidelines, Class III devices must also be premarket approved by the FDA.

Depending on an initial assessment of risk to the user by the medical device sponsor, devices are classified as significant risk (SR) or nonsignificant risk (NSR) devices. If a device is categorized as SR (most Class III devices), it is subject to clinical investigation under Investigational Device Exemption (IDE) regulations, and if it is an NSR device, it is subject to abbreviated IDE regulations.

An IDE is similar to an IND in that it allows an investigational device to be used in a clinical trial in order to collect safety and effectiveness data. If a device is classified as an SR device, clinical studies cannot begin until the IDE is approved by the FDA and by an IRB. The clinical trial sponsor and investigator must comply with the full list of IDE requirements in 21 C.F.R. Part 812.

If a device is classified as an NSR device, the sponsor does not need to submit an IDE to the FDA; however, the sponsor must follow the abbreviated requirements of an IDE relating to labeling, IRB approval, informed consent, monitoring, records and reports, and promotion. Rather than FDA approval, the IRB serves as the FDA's proxy for initial and continuing review of the NSR medical devices.

Many medical device companies are small companies with limited funding to conduct clinical research. However, this fact may or may not be a consideration for particular research sites that utilize device trials to meet the needs of their patients or clients.

Like drug trials, medical device studies are conducted to demonstrate the safety and effectiveness of the device before marketing, validating previous bench or animal testing. Pilot studies are usually conducted at single research sites with limited numbers of subjects to allow the sponsor to collect data on a series of patient outcomes that contribute to the selection of endpoints for pivotal trials. Conversely, pivotal trials are multicenter studies that focus on the safety and effectiveness of a device, which will eventually be used in support of the finished product's label. The studies are usually short in duration and may require dedicated space, especially if the investigational device is to be implanted.

SPONSOR SELECTION OF THE RESEARCH SITES

The cost of conducting clinical trials is a major part of the drug development process and budget, influencing sponsors to seek investigators and research sites that can recruit participants and conduct studies in an efficient, economical, ethical, and timely manner. Sponsors seek research sites that have adequate and qualified staff and have the facility/space and equipment resources necessary to properly conduct the clinical trial. In addition, sponsors consider the following:

- The research site's history of meeting recruitment goals in similar studies.
- Compliance with applicable Codes of Federal Regulations and International Conference on Harmonisation (ICH) Good Clinical Practices (GCPs).
- Investigator's interest in the clinical trial.
- A history of the research site being able to meet clinical trial deadlines.

It is also important to the sponsor that a research site is able to initiate a clinical trial in a timely manner. A research site that is not able to start a clinical trial quickly (IRB processes, contracting, and recruitment) delays the clinical trial, ultimately costing the sponsor time and money.

Financial motives are not the only consideration in a sponsor's selection of a site. Two key sponsor responsibilities according to 21 C.F.R. Part 312.50 (FDA IND General Responsibilities of Sponsors, 2015) are selecting qualified investigators to conduct clinical trials and ensuring that the clinical trial(s) are conducted in accordance with the general protocol and protocols contained in the IND application. These two responsibilities are also found in 21 C.F.R. Part 812.40 (FDA IDE General Responsibilities of sponsors, 2015). Additionally, the ICH E6 GCP Guidelines note that "investigators should be qualified by training and experience and should have adequate resources to properly conduct the trial for which the investigator is selected" (ICH, 1996, Section 5.6.1). Because the sponsor is the author/initiator of a clinical trial, it has the discretion to determine what qualifications, training, education, and experience are required of an investigator, the research team, and facility (U.S. Department of Health and Human Services, FDA, 2010, p. 6).

As part of its due diligence, sponsors examine multiple research site metrics to determine whether or not to select a research site. These metrics may include the following:

1. Adequate and qualified staff: Does the research site have adequate and qualified staff to conduct the clinical trial? This includes the principal investigator, clinical trial coordinator(s), and support staff needed for the trial (for example, lab technicians to analyze urine samples, a research pharmacist to mix investigational drug, or a data entry person). If the clinical trial includes procedures such as x-rays or biopsies, the sponsor must be assured that the research site has access to vendors qualified to perform the procedures and the ability to do the procedures in a timely manner. The sponsor is responsible for ensuring that the research staff are qualified and have not been barred by the FDA from conducting clinical trials as discussed in Chapter 1. This information assists the sponsor in avoiding the use of unqualified or barred investigators that could impact the reliability and validity of the clinical trial data. It also ensures that the investigator has staff that she or he can dedicate to the clinical trial to ensure that the clinical trial is given the attention needed to assure successful completion of the clinical trial. The ICH (1996) states that "The investigator should have available an adequate number of qualified staff and adequate facilities for the foreseen duration of the trial to conduct the trial properly and safely" (Section 4.2.3).

2. Adequate space and facilities: Does the research site have adequate space, required facilities, and equipment to properly conduct the clinical trial? Space includes safe and secure storage for the investigational product and clinical trial binders, a laboratory for tests such as urinalysis, appropriate exam rooms, waiting area, and appropriate office space. The research site must have the necessary general equipment, such as stethoscopes, electrocardiogram machines, and scales, to conduct the clinical trial procedures and appropriate office equipment such as computers, fax machines, and telephones. In some cases, the sponsor will require equipment calibration records and equipment maintenance records. Some trials require special facilities/equipment such as an area for IV infusions, beds, and the ability to accommodate a subject overnight or for other timeframes. Sponsors generally supply any specialized equipment required for a clinical trial, for example, infusion pumps or electronic clinical trials management software. In addition, the sponsor evaluates shared spaces (exam rooms, lab, etc.) and the number of studies, subjects, and visits that occur in the space to determine that the space is not overused and will be available as needed for clinical trial visits.
3. Clinical trial population: Does the investigator/research site have access to the clinical trial population? It does not matter how good or efficient a research site is if the research team cannot recruit subjects for the clinical trial. The research team must demonstrate its ability to recruit the required population in a timely fashion by having a solid recruitment plan. This plan might target patient data bases, referring physicians and clinics, or recruitment materials targeting the general population.
4. Support: Organization/department support is critical to the success of the investigator/research site and the clinical trial. The sponsor needs assurance that the key stakeholders, including upper management within the organization, are supportive of research and that the institution embraces a culture that includes clinical research. Without support, resources needed for research may not be available or limited, and recruitment of subjects may not be an organizational priority.
5. Current research site obligations: Research sites that conduct multiple studies run the risk of having competitive studies that require the same clinical trial population. The sponsor needs assurance that the research team (including the investigator) can effectively manage its current work load as well as add another clinical trial. A sponsor will be hesitant to place a clinical trial at a research site that has current active studies that are recruiting the same population as the new clinical trial—this places the trial in a competitive position against other trials. The research team may overcome this obstacle if it has a large patient population to draw from and if the clinical trial's inclusion and exclusion criteria, although similar, have differences that would allow the screening of subjects who did not meet the criteria from other studies.
6. Recruitment history: An investigator's subject recruitment history is indicative of the research site's commitment to meet the agreed upon number

of subjects for other studies. Investigator/research sites that have difficulty meeting recruitment goals are less likely to be included in the final research site selection. ICH notes that "The investigator should be able to demonstrate (for example, based on retrospective data) a potential for recruiting the required number of suitable subjects within the agreed recruitment period" (ICH, 1996, Section 4.2.1).

7. FDA Inspection history: Regulatory compliance is a major factor in the conduct of clinical trials. FDA Form 483 Inspectional Observations, FDA warnings, and audits are all cause for concern when a sponsor is evaluating research sites. Although it has become more common for research sites to receive Inspectional Observations after an FDA inspection, a sponsor is interested in the numbers and types of observations and the investigator's responses and resolutions of the identified deviations. Warning letters from the FDA indicate serious problems at a research site. A warning letter is a red flag to a sponsor but may not eliminate an investigator for consideration if she or he is able to demonstrate that the issues were appropriately addressed and resolved and show there have been no similar problems since receipt of the warning letter.
8. Clinical trial oversight: Investigator oversight of personnel and clinical trial conduct is essential and mandatory for clinical trials. Both the sponsor and the FDA want to be assured that an investigator will be actively involved in all aspects of the clinical trial rather than acting as a figurehead only. She or he should be on site and involved in all aspects of the trial. Delegating clinical trial responsibilities to qualified personnel does not remove the investigator's oversight responsibility.
9. Current training and education: Research teams often include investigators, coordinators, nurses, laboratory personnel, data managers, pharmacists, assistants, and others. As such, the need to track licensure, certification, scope of practice, and continuing education is very important to ensure that the investigator and his or her research team are qualified by training and education to fulfill their roles during a clinical trial. Research site policies should define the minimum training and education required of the various clinical trial roles. New employees should meet at least the minimum standard requirements pertinent to their role on the clinical trial. Additionally, many research sites require a research-related certification for clinical trial staff.

   A professional competency certification demonstrates that the holder of the certification has met minimum standards within the profession. (Professional organizations that provide research-related certifications are found in Appendix C.) Credentialing logs and checklists can assist investigators in maintaining documentation for review by the sponsor or regulatory agencies.

   Research site policies should also define staff training and education in research. All training should be documented in personnel or training files, which should be updated on a regular basis. Accurate and complete training records are important to demonstrate to the sponsor that the investigator and research site personnel are appropriately trained and qualified. The ICH GCP addresses this requirement stating, "Each individual involved in conducting a trial should be qualified by education, training, and experience to perform his or her respective task(s). The investigator should ensure that all persons assisting with the trial are adequately informed about the protocol, the investigational product(s), and their trial-related duties and functions" (ICH, 1996, Section 2.8).

10. IRB: A lengthy IRB process may delay the start of a clinical trial, which, in turn, may delay the entire clinical trial across several research sites. This may be a major factor when a sponsor selects an investigator/research site. Given the opportunity, investigators should consider IRB turnaround time when they review their options for selecting an IRB. If the investigator is able to use the same central IRB that approved the protocol for the sponsor, it will expedite the process.
11. Barriers and limitations: It is important to identify any potential barriers/limitations that an investigator might foresee in conducting a clinical trial. The investigator needs to be forthright with sponsors and have plans in place to address any identified barriers and limitations. For example, a research site might not currently have enough qualified coordinators to add another clinical trial to its current offerings. In this case, the investigator might provide a timeline and training schedule to a sponsor to demonstrate that she or he is committed to its clinical trial; the research site may have several coordinators who are currently being trained and will be appropriately qualified by the time the clinical trial starts. Remember, it is the also the investigator/research site's responsibility to ensure the welfare and safety of the clinical trial subjects. Failing to address or disclose possible limitations could ultimately impact the safety of clinical trial subjects.

CONFIDENTIAL DISCLOSURE AGREEMENTS

The discovery, development, and pre-IND costs and obligations incurred by a sponsor prior to placement of a clinical trial at a research site are enormous. It is no surprise that sponsors wish to maintain a marketing advantage against their competitors by requiring all parties familiar with their investigational products to maintain confidentiality. Sponsors consider their investigational products to be intellectual property and require investigators to sign a CDA that establishes the terms and conditions for release of the sponsor's clinical trial protocol, investigator brochure, informed consent forms, and other clinical-trial-related materials.

Because the CDA is a legal document, the investigator (and research team) is subject to legal action if there is a breach of confidentiality. Breaches can happen unintentionally in the process of collaboration between investigators, in the recruitment of subinvestigators, in working with referring physicians, and in many other ways. It is important that the research team has consistent processes in place for ensuring standardized communication involving proprietary information at the research site.

This can be accomplished by clearly marking confidential information (with watermarks or other visible reminders) and by following a standard operating process for handling proprietary information.

CLINICAL TRIAL FEASIBILITY

In this section, we discuss protocol feasibility, the idea that a clinical trial protocol is appropriate and "feasible" to be conducted by a specific investigator at a particular research site. For a clinical trial protocol to be feasible, it must meet the regulatory obligations of the sponsor (as reviewed in Chapter 1 and previously in this chapter) and it must fit the resources, capabilities, and qualifications of an investigator and the research site. Additionally, the clinical trial should be scientifically and ethically sound from the perspectives of both the sponsor and the investigator.

Although it is logical to assume that matching a clinical trial protocol to an appropriate investigator will ensure successful conduct of a clinical trial, there are many factors that impact feasibility assessments and strategies. In this section, we review the regulatory foundations of clinical trial feasibility and then explore the role of the investigator in determining protocol feasibility.

FOUNDATIONS OF FEASIBILITY

Although federal regulations and GCP guidelines do not specifically mention "clinical trial feasibility," they emphasize the roles of the sponsor and investigator in ensuring that the investigator is qualified to conduct a clinical trial and that the investigator commits to conducting a clinical trial according to the protocol. As mentioned in Chapter 1, key responsibilities for investigators include compliance with the IRB-approved protocol and the obligations found in Part 9 of the "Statement of the Investigator" for studies compliant with 21 C.F.R. Part 312. Similar obligations are found in 21 C.F.R. Part 812 in the "investigator letter." Along these same lines, ICH E6 Guidance notes in Parts 4.1 and 4.2 that the investigator must be qualified and that she or he should have the potential for recruiting the required number of suitable subjects, have sufficient time to conduct the clinical trial, and have adequate staff and facilities (ICH, 1996, pp. 12–13). To summarize, federal regulations and ICH GCP specify that the investigator should be qualified and have the resources needed to conduct the clinical trial.

Considered together, the above regulations form the foundation and purpose for reviewing studies to determine their "fit" for the investigator and his or her research site prior to clinical trial placement at the site.

Not every clinical trial is a good fit for an investigator and his or her research site, regardless of the investigator's qualifications. An investigator should review a clinical trial from a scientific, regulatory, ethical, and resource/logistic perspective. These perspectives will not only aid the investigator in determining clinical trial feasibility, it will also assist him or her in determining whether the clinical trial agreement and budget will sufficiently meet the investigator and research site needs.

Scientific Review

Ideally, a clinical trial protocol and its supporting documentation arrive at the research site ready for activation. Yet, there are times when the clinical trial is still in draft version, contains missing information or documentation, and/or is not scientifically sound. As such, it is important that the research site has a process in place for the intake and quality check of the documents, which might include the protocol, informed consent form, Health Insurance Portability and Accountability Act form, investigator brochure, recruitment materials, and/or contract. The first step in review of the clinical trial should include an inventory of documents using a research site checklist. Additional intake steps should include comparison of the clinical trial documents to ensure that the titles, versions, protocol numbers, page numbers, and other elements are consistent across all documents. Once checked, the documents should also be verified for consistent inclusion/exclusion criteria, calendars, doses, data to be collected, event reporting, and clinical trial termination. These editorial/quality assurance steps should be conducted prior to scientific review by the investigator and the research team.

Reviewing a clinical trial for scientific merit should be part of the scientific feasibility review of a clinical trial. The investigator, who has the background and qualifications to conduct the clinical trial, should evaluate the clinical trial to determine if it may have direct benefit to the population or to the field of discovery itself. Does it fit into the scope of the investigator's practice or area of expertise? The investigator should also evaluate the scientific methodology of the clinical trial. Among other criteria, she or he should review the clinical trial for design relative to the disease being studied, objectives and expected outcomes, data to be collected appropriate to investigational product safety and efficacy, and inclusion/exclusion criteria realistic for the disease under investigation. As the individual responsible for the conduct of the clinical trial at the research site, the investigator should be able to justify the protocol design and accompanying risks and benefits to the IRB and to clinical trial subjects.

Regulatory Review

It is rare that a clinical trial is not compliant with regulations. However, local and/or state laws may require compliance in a subject area of clinical trials unknown or unfamiliar to the sponsor. In this case, the regulatory concern needs to be addressed prior to investigator/research site participation.

It is also unusual that a clinical trial protocol does not include the items required by IND regulations, 21 C.F.R. Part 312.23 (a)(6)(iii). Even so, it is good practice for investigators and their clinical trial personnel to utilize a checklist when reviewing a clinical trial to ensure that the following elements are contained in the protocol.

- Statement of objectives and purpose of the clinical trial;
- Criteria for subject selection;
- Description of the design of the clinical trial;
- Method for determining dose(s) and duration of exposure;

- Description of observations/measurements to fulfill objectives; and
- Description of clinical procedures to monitor effects and minimize risks (IND Content, 2015).

Ethical Review

Ethical review of a clinical trial is usually associated with IRB processes, but it is also the responsibility of the investigator to review the clinical trial prior to IRB review for compliance with the Belmont Principles of respect for persons, justice, and beneficence. In particular, she or he should identify informed consent and autonomy criteria, equitable subject selection criteria, and risk–benefit balance relating to the research site's clinical trial population. Additionally, the investigator should weigh the ethical considerations with the scientific merit of the clinical trial. Unfortunately, scientific rigor and ethical conduct may be a difficult balance with some high-risk studies. Yet, the investigator should be aware of clinical trial strengths and weaknesses as he or she contemplates the research site's "fitness" to conduct the clinical trial.

Resource Review

The scientific, regulatory, and ethical reviews of a clinical trial are foundational in determining whether the investigator has an interest in conducting a particular clinical trial at his or her research site. Conversely, a review of the clinical trial to determine resource feasibility is directly related to conduct of the clinical trial at the research site. The investigator must have adequate facilities, equipment, supplies, personnel, and time commitment to manage and conduct the clinical trial.

The investigator and his or her research team should carefully review the clinical trial documents to determine whether the clinical trial population and resources are available at the research site and whether there are unique requirements that may require special accommodation. Some of the questions that should be considered include the following:

- Does the research site have enough space to conduct the clinical trial?
- Does the research site have the appropriate (calibrated) equipment?
- Does the clinical trial require specific supplies, and will they be supplied by the sponsor?
- Does the clinical trial require that specific clinical trial-related tasks be conducted under specialty (professional) licensures?
- Does the clinical trial require specific time commitments from the investigator?
- Does the investigator have competing time commitments?

Obviously, these questions are limited in scope and do not address some of the intricacies that should be considered, especially in very complex studies. As such, a typical research site feasibility checklist is available in Appendix B.

As a process, feasibility review lends itself very well to an SOP. When written in a step-by-step manner, the SOP can ensure that all studies are reviewed using the same criteria and that the process is consistent with different personnel and investigators. This will save time and allow the research site to have a single, specific review process. A sample SOP is found in Appendix B.

IRB APPLICATIONS/SUBMISSIONS

Because a clinical trial cannot begin until an IRB has approved it, pretrial site preparation should, at the very least, include a process or SOP for IRB submissions that meet the investigator's responsibilities for IRB approval. It is his or her responsibility to ensure that "an IRB that complies with the requirements of 21 C.F.R. Part 56 will be responsible for the initial and continuing review and approval of the clinical investigation" (U.S. Department of Health, FDA, 2013, p. 2). This means that the investigator must choose an IRB that meets the regulatory requirements for human subject protections when preparing for the clinical trial.

In many organizations, it is a policy that investigators work with the organization's internal IRB. In doing so, investigators trust that the organization complies with Title 21 regulations and that it maintains a Federal Wide Assurance (FWA) through the Office of Human Research Protections.

When not required by the investigator's organization, investigators engage with the IRB of their choice. In some cases, an investigator chooses a specific IRB, and in other cases, the investigator may choose an IRB selected by the sponsor. Regardless of choice, the investigator is responsible for supplying the IRB with details of the clinical trial, including information in regard to the research team, recruitment strategies, timelines, and more, for review by members of the IRB. Most IRBs require this information to be submitted through an electronic process that tracks clinical trial intake, processing, review, timelines, and types of approval.

Institutional review boards have their own processes and review requirements. However, the most common review categories are full board review, expedited review, and exempt category (protocol is exempt from review). Full board review means that the clinical trial and supporting materials are reviewed by the entire committee; expedited review means that the clinical trial presents minimal risk and it does not require the review of the full board (only assigned members); and exempt means that the clinical trial falls into a category defined by regulations to be exempt from review. Most industry-sponsored research studies require full board review due to the nature of investigational product risks. However, recruiting materials or advertisements for a clinical trial are usually reviewed through expedited review.

At the research site, the process of submitting an IRB application can be confusing and overwhelming, but it lends itself to the use of standard operating processes and considerations in defining who is responsible for each aspect of the process and how it is to be accomplished. Although individual clinical trial protocols are not alike, the overall processes should be consistent.

The IRB application requires the applicant to be thoroughly familiar with the protocol, the investigator brochure, sponsor documentation, financial disclosure, clinical trial calendars, risks/benefits to subjects, special populations, and informed consent. Consequently, the application process should include the investigator, who may need to consult with the sponsor on various aspects of the IRB application to clarify information.

In many cases, the sponsor will include an informed consent template, case report forms, advertising materials, and other forms with the clinical trial protocol. Some IRBs may require that an investigator (or his or her delegate) alter the sponsor's informed consent form template to meet the IRB's standard template, and the sponsor may want to review the template before it is submitted to the IRB. Additionally, some IRBs may not accept draft protocols or drafts of any of the clinical trial materials, so it is very important that the research team understands the policies and procedures of individual IRBs prior to submission.

The IRB process is an important aspect of the clinical trial, and it is also an expensive part of the process. Institutional review boards charge the research site (or the sponsor directly) to review a clinical trial, and the investigator/research site expends tremendous man-hours on IRB submission. Although the IRB application process is considered a pretrial operation, the research team should also have contracts in place with the clinical trial sponsor to ensure that the time expended on the process is not uncompensated or wasted.

ESSENTIAL DOCUMENTS

Documentation is an essential element of clinical trials, providing a mechanism of accountability for everyone involved. The ICH E6 GCP Guidance labels these documents as "essential documents" and defines them as "those documents that individually and collectively permit evaluation of the conduct of a trial and the quality of the data produced" (p. 50).

During pretrial preparation, the ICH suggests that the following documents are in place at the research site:

- Investigator's brochure
- Signed protocol and amendments and sample case report form
- Informed consent form, written information, advertisements for subject recruitment
- Financial aspects of the trial
- Insurance statement (where required)
- Signed contract among involved parties
- IRB approval (dated/documented)
- IRB composition (who is on the Board)
- Regulatory authorization (where required)
- Curriculum vitae (CV) and/or other relevant documents of investigators and other research site personnel

- Normal values/ranges for medical/technical procedures
- Instructions for handling of investigational products
- Shipping records for investigational products
- Decoding procedures for blinded trials
- Trial initiation monitoring report (ICH, 1996, pp. 51–54)

Other essential documents that should be completed during pretrial preparation include documentation of the current medical license number for the investigator, the Statement of the Investigator (FDA Form 1572), financial disclosure information, the research site information sheet (or research site demographics form), and clinical trial personnel CV and training records. There may be additional forms, logs, checklists, and/or information required by the sponsor or through the research site's SOPs.

In Chapter 3, we will provide more detail involving essential document preparation and will explore the requirements of source documentation and case report form preparation.

A VIEW FROM INDIA

Investigational New Drugs

The application for permission to conduct a clinical trial for an IND in India must be made to the Drugs Controller General of India (DGCI). The phases as described in this chapter have similar requirements in terms of the subjects (type and number) as in India. However, the objectives of the study phases are slightly different from those in the United States, as described in Schedule Y:

- Phase I (Human Pharmacology): Phase I studies are conducted to find the maximum tolerated dose, pharmacokinetic and pharmacodynamic effect, and safety profile of the IND. The potential therapeutic activity may be the secondary objective.
- Phase II (Exploratory trials): The primary objective of the phase II studies is the efficacy along with further evaluation of safety and pharmacokinetics of the IND.
- Phase III (Confirmatory trials): The objectives of phase III studies are to demonstrate efficacy and safety of the IND in the patient population.
- Phase IV (Postmarketing trials): Phase IV trials are conducted in the form of postmarketing surveillance after the IND is approved for marketing. Its objective is to assess the therapeutic value, the safety profile, and the treatment strategies used (Government of India, Ministry of Health & Family Welfare [MOHFW], 2013, pp. 734–736).

For new drugs discovered in India, clinical trials are required to be initiated from phase I within India. For new drugs discovered outside India, phase I data should be generated outside India and submitted along with the application to DCGI to conduct subsequent phases of clinical trials in India. For such applications, the licensing authority may require the sponsor to repeat phase I studies in India before subsequent phases can be conducted. Phase III studies must be conducted in India, if the new drug is intended to be marketed in India (Government of India, MOHFW, Central Drugs Standard Control Organization [CDSCO], 2011, pp. 7–8).

Medical Devices

Currently, the application for permission to conduct a clinical trial for new medical devices in India must be made to the DGCI. Only "notified" medical devices are regulated in India. (The notified medical devices are listed in Chapter 9.) New medical devices are those for which devices with the same indication, intended use, material of construction, and design characteristics are not (already) registered in India. The CDSCO has setup a Medical Device Advisory Committee to advise the DCGI in matters related to clinical trials and regulatory approval of new medical devices.

Medical devices are classified into two categories depending on the risk involved:

1. Noncritical devices—Investigational medical devices that do not pose a significant risk to the patients, for example, thermometer and blood pressure apparatus.
2. Critical devices—Investigational medical devices that pose a serious risk to the health, safety, or welfare of the patients, for example, pacemakers, implants, and internal catheters (Indian Council of Medical Research New Delhi, 2006, p. 49).

The 2016 copy of the draft of Medical Devices Rules can be found at: http://cdsco.nic.in/writereaddata/Draft_Medical%20Devices%20Rules%202016.pdf.

Selection of Research Sites

The selection of research sites in India is similar to the process in the United States. Additionally, research sites must have the facilities to conduct the consent process using audio video equipment, as this is required in India.

Accreditation of Research Sites

As recommended by the Ranjit Roy Chaudhary expert committee, 2013, clinical trials must be conducted only at accredited clinical trial sites by accredited investigators under the oversight of an accredited IRB/independent ethics committee (IEC). The National Accreditation Board for Hospitals and Healthcare Providers, Quality Control of India has set up accreditation standards for ethics committee, investigators, and clinical research sites.

Confidential Disclosure Agreements

The objective of the CDA has similar intent and purpose in India as it is in the United States.

IRB/IEC Submissions

The objectives and responsibilities of the ethics committees, as well as the documents required for submission, are similar in India and the United States. In India, the informed consent form translation to regional language and back-translation to English is also required for submission to the ethics committee.

Essential Documents

Essential documents required in India are similar to those required in the United States. Other essential documents required in India include the investigator's current registration with the Medical Council of India (MCI) and the Investigator's Undertaking (discussed in Chapter 1).

CHAPTER REVIEW

In this chapter, we examined the pretrial preparation at the research site, focusing on the different considerations involved with investigational drug and device clinical trials, research site selection, confidentiality, clinical trial feasibility, IRB review, and essential documents. Additionally, we summarized some of the unique elements of India's pretrial preparation requirements.

APPLY YOUR KNOWLEDGE

1. After signing the CDA, Dr. Trail and her research team begin to review the clinical trial for feasibility. During review, the research team notes that the clinical trial IND is still pending within the 30-day approval timeframe. What concerns should Dr. Trail have in regard to this situation?
2. PharmaXYZ has placed five studies with Dr. Marsh and his research team. The studies have gone well and have been successfully completed. PharmaXYZ is eager to place another arthritis clinical trial with Dr. Marsh and his research team. However, Dr. Marsh has reviewed the clinical trial and determined that it is almost identical to the last two studies, and it accesses the same population. Which of the four review areas do you think is most relevant in his review of the clinical trial, scientific, regulatory, ethical, or resource? Why?
3. Dr. Callen is reviewing a clinical trial to determine whether he is interested in becoming an investigator for the clinical trial. He is extremely interested in the science behind the clinical trial and believes that it will offer a great contribution to generalizable knowledge in the discipline. As he reviews the investigator brochure, he notices that many of the animals in the preclinical trials suffered severe toxicities and died shortly after administration of the investigational drug. He returns to the protocol and the informed consent form to determine if any of these reactions are considered risks in the clinical trial but does not find the information mentioned. Is this of concern? Why or why not? Next steps?

REFERENCES

FDA IDE General Responsibilities of Sponsors, 21 C.F.R. § 812.40 (2015).

FDA IND General Responsibilities of Sponsors, 21 C.F.R. § 312.50 (2015).

Government of India, Ministry of Health & Family Welfare, Central Drugs Standard Control Organization. (2011). Draft guidance on approval of clinical trials & new drugs. Retrieved from http://www.cdsco.nic.in/writereaddata/Guidance_for_New_Drug_Approval-23.07.2011.pdf

Government of India, Ministry of Health & Family Welfare. (2013). The drugs and Cosmetics Act, 1940 and rules, 1945 Schedule Y. Retrieved from http://www.mohfw.nic.in/WriteReadData/l892s/43503435431421382269.pdf

Indian Council of Medical Research New Delhi. (2006). *Ethical guidelines for biomedical research on human participants*. Retrieved from http://www.icmr.nic.in/ethical_guidelines.pdf

International Conference on Harmonisation. (1996). *Guideline for good clinical practice E6(R1)*. Retrieved from http://www.fda.gov/downloads/Drugs/GuidanceComplianceRegulatoryInformation/Guidances/ucm073122.pdf

U.S. Department of Health, Food and Drug Administration. (2013). *Statement of investigator*. Retrieved from http://www.fda.gov/downloads/AboutFDA/ReportsManualsForms/Forms/UCM074728.pdf

U.S. Department of Health and Human Services, Food and Drug Administration. (2010). *Information sheet guidance for sponsors, clinical investigators, and IRBs: Frequently asked questions—Statement of investigator*. Retrieved from http://www.fda.gov/downloads/RegulatoryInformation/Guidances/UCM214282.pdf

3 Sponsor, Site, and Study Start-Up

The cost of developing and bringing a new drug to market is extremely high. In a 2014 study by the Center for the Study of Drug Development at Tufts University, the cost of bringing a single drug to market was estimated at US $2.558 billion (DiMasi, Grabowski, & Hansen, 2016, p. 25). Of that amount, approximately US $1.460 billion was attributed to the cost of conducting clinical trials (DiMasi et al., 2016, p. 25). With this amount of money at stake, sponsors are motivated to collaborate with research sites that are prepared and ready to initiate a clinical trial in a timely manner once the clinical trial protocol and related documents are approved by the institutional review board (IRB).

This chapter reviews the various steps and activities that need to occur at the research site prior to starting the clinical trial. We review the forms and documents required of the investigator/research site to start a clinical trial and describe the Statement of the Investigator in detail because it is the investigator's commitment to conduct the clinical trial in accordance with all applicable laws and regulations.

RESEARCH SITE START-UP RESPONSIBILITIES

The research team has obligations and tasks that must be completed before a clinical trial is started. The research team must provide the sponsor with required documents and complete the necessary research site set-up activities before clinical trial start-up. These include the following:

1. Complete and submit Form FDA 1572.
2. Submit and receive approval of the research site-specific informed consent from an appropriate IRB. A research site-specific consent is the approved general consent form that includes the research site's information. Research site information includes the research site name, address, contact information, investigator name and contact information, and IRB contact information.
3. Ensure and document that all staff involved in the conduct of the clinical trial have completed necessary clinical trial specific training. The investigator is responsible for ensuring that his or her research team has reviewed and understands the protocol, consent form, and other clinical trial materials. The investigator may delegate the tracking of this information to a manager or supervisor.
4. Provide the sponsor with investigator(s), subinvestigator, and clinical trial physician curriculum vitae (CVs), medical licenses, and proof of insurance. Sponsors may also require resumes and certifications of other research staff.
5. Complete a Delegation of Authority form. The investigator, although responsible for the conduct of the clinical trial, is able to delegate clinical

trial tasks and responsibilities to staff who are qualified to carry out the tasks/responsibilities. For example, the investigator will assign responsibilities to the clinical trial coordinator, research nurse, laboratory technicians, and other research staff. A clinical trial coordinator may have the authority to conduct the informed consent, take vitals, and conduct clinical trial visits, including procedures such as electrocardiograms and blood draws. A research nurse may have the authority to conduct physical exams and to perform infusions and other procedures required for the clinical trial. Noted in International Conference on Harmonisation (ICH, 1996, Section 4.1.5), "The investigator should maintain a list of appropriately qualified persons to whom the investigator has delegated significant trial-related duties."

6. Document the receipt of supplies, materials, and equipment. This includes the inventory of all materials received and appropriately storing the clinical trial supplies.
7. Document the receipt of the investigational product (drug or device), inventory investigational product, and store according to instructions.
8. Ensure the electronic data capture (EDC) system required by the sponsor is installed and working and that staff who will be entering or signing off on data are appropriately trained in use of the system. Staff should also be trained and familiar with 21 C.F.R. 11: Electronic Records and Signatures. All staff using the EDC must provide a written signature and appropriate identification for access and use of the electronic data system.
9. Appropriate research site staff must review, negotiate, and approve the Clinical Trial Agreement.
10. Appropriate research site staff must review, negotiate, and approve the clinical trial budget.
11. The research team should create a recruitment plan that includes the identification of potential participants for the clinical trial. The plan must include activities that the research team will conduct to recruit clinical trial subjects (for example, chart reviews, mailings, and radio or television ads), a recruitment schedule, and any recruitment materials that will be used. All recruitment materials must be approved by an appropriate IRB prior to use. (A Research Site Start-up Checklist for research sites is found in Appendix B.) Figure 3.1 shows the flow of activities at the clinical site prior to study start-up.

Investigator Forms

As a part of sponsor's Investigational New Drug (IND) or Investigational Device Exemption (IDE), an investigator must agree to conduct studies in compliance with FDA regulations by signing a "Statement of the Investigator," which is handled in two different manners, depending on whether the clinical trial involves drugs or medical devices. Regulations for INDs require the investigator to complete Form FDA 1572. IDE regulations require the investigator to sign a letter of agreement supplied by the sponsor that he or she will comply with the investigator obligations under 21 C.F.R. 812.43(c). The sponsor must collect these forms and inform the FDA that signed agreement forms have been received and are on file.

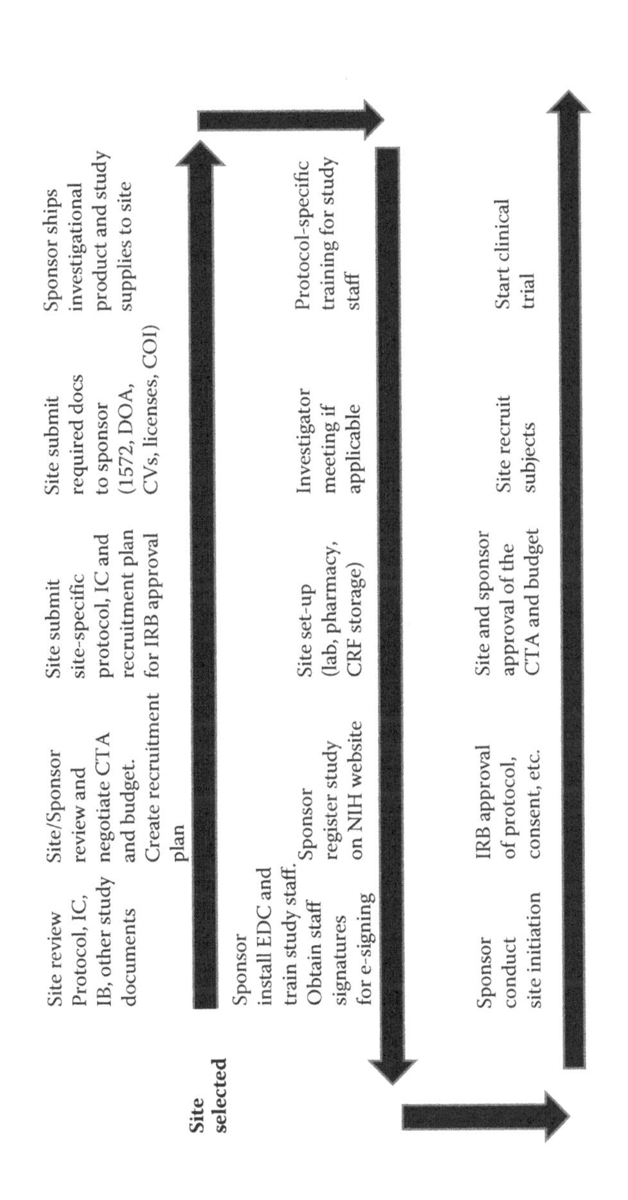

FIGURE 3.1 Research site start-up.

Form FDA 1572 is a legal agreement between the investigator and the FDA for drug studies. By signing Form FDA 1572, the investigator accepts responsibility for conducting the clinical trial in accordance with FDA regulations and guidelines. The investigator's signature on the Form FDA 1572 indicates his or her agreement to provide direct oversight of the clinical trial and to accept ultimate responsibility for the conduct of the clinical trial at the research site. According to 21 C.F.R. 312.53(c), the sponsor must obtain the following information on Form FDA 1572 before permitting the investigator to begin participation in a clinical trial. Table 3.1 provides a summary of the investigator's obligations included in Form FDA 1572. Table 3.2 provides a summary of the investigator's agreement for investigational devices.

TABLE 3.1
Statement of the Investigator Summary

Obligation	Description
Provide contact information	To include name and address of the investigator.
Clinical trial information	Name and code number of the protocol.
Research facility	Name and address of all facilities where the research will be conducted.
Clinical laboratory	Name and address of any clinical laboratory facilities to be used in the clinical trial.
IRB	Name and address of the IRB responsible for the review and approval of the clinical trial.
Commitment to:	• Conduct the clinical trial in accordance with protocol. • Only make changes after notifying the IRB and sponsor, except when necessary to protect the subject. • Personally conduct or supervise the investigation. • Inform potential subjects that the drug being used is investigational. • Ensure requirements relating to obtaining informed consent and IRB review and approval are met. • Report adverse events that occur during the clinical trial. • Read and understand the IB, including the potential risks and side effects of the drug. • Ensure that all associates and employees assisting in the conduct of the clinical trial are informed about their obligations.
Institutional review	Ensure that the clinical trial will comply with the review requirements under 21 CFR 56.
Sub-investigators	Provide a list of the names of sub-investigators who will be assisting in the conduct of the clinical trial.

TABLE 3.2
Investigator Agreement: IDE

Obligation	Description
Investigator qualifications	Provide current curriculum vitae.
Investigator statement	Attesting to relevant experience, including dates and location, extent and type of experience.
Clinical trial termination	An explanation of the circumstances that led to the termination of any clinical trial the investigator was involved in.
Commitment to:	Conduct the investigation in accordance with the agreement, investigational plan, IDE and other applicable FDA regulations and conditions of approval imposed by the reviewing IRB. • Supervise all testing of the device involving human subjects. • Ensure that the requirements for obtaining informed consent are met.
Financial disclosure	Submit a complete and accurate disclosure statement as required under 21 CFR 54.

Other Information Required of the Investigator

The sponsor is required by the FDA to obtain the following documents from the investigators and keep them on file for FDA review. A copy of these documents should be given to the sponsor and a copy kept in the research site's regulatory binder.

- A current CV or other statement of qualifications of the investigator and subinvestigators showing the education, training, and experience that qualifies the investigator as an expert in the clinical investigation of the drug or device for the use under investigation.
- A copy of the investigator and subinvestigators' current medical licenses to ensure that they are able to conduct the clinical trial, diagnose, treat, and prescribe medications. If the clinical trial involves the use of narcotics, the sponsor must collect a copy of US Drug Enforcement Agency (DEA) licenses of the investigator and coinvestigators. The DEA license is what allows the physician to prescribe controlled substances/narcotics as these can be addictive.
- Financial disclosure information. Many physicians, including investigators, have a financial association with pharmaceutical companies. The physician may act as a consultant for the company, may be on a speaker's bureau, and have ownership interest, stock options, or other financial interest in a pharmaceutical company. Disclosure of such information is necessary to ensure that the financial interests and arrangements of the investigator with a sponsor company do not affect the reliability of the clinical trial data. 21 C.F.R. Part 54 requires reporting these relationships if the monetary value is equal

to or above $50,000. The sponsor must obtain sufficient accurate financial information from the investigator, subinvestigators, and research staff that will allow the sponsor to submit complete and accurate certification and disclosure statements required under 21 C.F.R. Part 54. In addition, the sponsor must obtain a commitment from the investigator to promptly update this information if any relevant changes occur during the course of the investigation and for 1 year following the completion of the clinical trial.

- The sponsor will also require the research team members to sign a confidentiality agreement. This agreement assures the sponsor that the research team who has access to proprietary and other confidential information will not share that information with another party who has not signed the agreement. This agreement also addresses access to confidential information by role on the clinical trial. For example, an investigator will have access to all clinical trial information such as drug information, adverse events (AEs), budget, and IRB reports. A research nurse may only have access to subject information for procedures that he or she performs. A pharmacist will have access to investigational product information and distribution.

Other Clinical Trial Documents

The investigator is responsible for maintaining all clinical trial records and ensuring that the data are accurate and complete. In addition to the essential documents described previously, there are other documents that must be completed and maintained. Review of these documents enables a monitor and an inspector/auditor to review the records and reconstruct the trial from day of initiation to determine if the trial was conducted in compliance with the protocol and all applicable regulations and guidelines. These include the following:

Regulatory binder: Each research site must maintain a regulatory binder that contains the essential documents, as mentioned in Chapter 2. The ICH (1996, Section 8.1) defines essential documents as "those documents that individually and collectively permit evaluation of the conduct of a trial and the quality of the data produced. These documents serve to demonstrate the compliance of the investigator, sponsor, and monitor with the standards of good clinical practice (GCP) and with all applicable regulatory requirements." The regulatory binder should be reviewed for accuracy and completeness by the monitor at each research site visit.

An example of the regulatory binder table of contents follows:

- Protocol—this section contains the original and all amended protocols.
- Informed consent—this section contains the original and all amended informed consents.
- IRB—this section contains information on the IRB of record, including contact information and members, IRB approvals, continuing review reports, and any communications with the IRB.
- Monitor logs—this section contains the monitoring log recording each visit, the monitor's report, and the site's follow-up report from each monitor visit.

- Delegation of authority—this section contains the delegation of authority and any updates or changes that occur during the conduct of the clinical trial.
- Personnel—this section contains research site personnel signature logs for electronic signatures, training and education logs, investigator and key staff CVs, financial disclosures, confidentiality agreements, and the Statement of the Investigator.
- Screening and enrollment log—this section contains the screening log and the enrollment log for the clinical trial.
- Clinical trial subjects—this section contains the subject visit tracking log and a list of the subjects' identification numbers.
- Medwatch or Suspected Unexpected Serious Adverse Reaction (SUSAR)—Reports that the research site receives on unexpected or serious AEs (SAEs) that have occurred at other sites in a multicenter study are contained in this section.
- Advertising/Recruitment—this section contains the research site's recruitment plan and advertising and recruitment materials.
- Temperature logs—this section contains temperature logs for investigational product that requires storage at a certain temperature.
- Laboratory—this section contains local lab certificates and range references.
- Deviations and violations—this section contains all of the research site deviation and violation reports during the conduct of the clinical trial.
- AEs/SAEs—this section contains a copy of adverse, unexpected, and serious AEs that occur at the research site during the clinical trial.

The regulatory binder should be reviewed by the monitor at each research site visit for completeness and accuracy.

Case report forms (CRFs): Clinical trial data are collected on a CRF. Data from the subject visits will be used by the sponsor to prove safety and efficacy of the investigational product. The FDA will review these data to approve a new drug for market.

The CRF may be in paper or electronic format. Case report forms are used at each visit to collect clinical trial outcomes data. Data collected on the CRF include the results from the various visit procedures and tests, for example, vitals data (heart rate, blood pressure, weight, etc.), results from blood draws and other tests, date of visit, times, and who collected the data. The CRFs are generally provided to the research site by the sponsor. If the data are captured on a paper CRF, they are usually then entered into an EDC system. In this case, it is important to compare the data on the paper CRF with the data entered into the EDC to ensure that data have been entered correctly before saving the data. Some systems require double entry of data to check for mismatched entries.

Source documents: The ICH E6 document, Section 1.52 (Appendix A), defines source documents as "Original documents, data and records (for example, hospital records, clinical and office charts, laboratory notes, memoranda, subjects' diaries of evaluation checklists, pharmacy dispensing records, recorded data from automated instruments, copies or transcriptions certified after verification as being accurate and complete, microfiches, photographic negatives, microfilm or magnetic media, x-rays, subject files, and records kept at the pharmacy, at the laboratories, and at medico-technical departments involved in the clinical trial)." Data from source documents may be required for the CRFs.

The source document is the original document on which the data are collected. It can be paper or an electronic health record. Some research sites might photocopy the CRF and use it as a source document to collect subject visit data. The data are then transferred to the CRF or electronic CRF. However, FDA guidance encourages the use of EDC for source documents as stated in a recent guidance document. "In an effort to streamline and modernize clinical investigations this guidance promotes capturing source data in electronic form, and it is intended to assist in ensuring the reliability, quality, integrity, and traceability of data from electronic source to electronic regulatory submission" (U.S. FDA, 2013). Technology will continue to advance. More sponsors are transitioning to paperless trials and this trend will continue to grow.

Investigator Meeting

Typically, the sponsor will host an investigator meeting prior to study start. Investigators and coordinators from each of the study sites are invited and encouraged to attend. During this meeting, the protocol, consent, investigator brochure, and other study documents are discussed and reviewed. Coordinators are trained on completion of CRFs and any surveys or questionnaires used in the study. In addition, training for protocol specific procedures and use of the EDC system occur at this time. Other topics covered might include AE and SAE reporting, study recruitment, screening of potential subjects, randomization, study monitoring, resolution of queries, GCPs, industry best practices, and other study-related items.

A VIEW FROM INDIA

The steps, activities, documents, and responsibilities for site initiation are similar in India to those in the United States. As mentioned in Chapter 1, investigators in India must sign a document similar in format, purpose, and content to U.S. Form FDA 1572, called the Investigator's Undertaking.

The (Indian) Research Site Master File is equivalent to the regulatory binder found in the United States, as it contains similar essential documents.

A few key differences between India and the United States include the following:

1. In India, it is required that patient information sheets and informed consent forms are translated into the language spoken by the local population being recruited by the research site. Back-translation to English of these documents is recommended and may be required by some IRB/independent ethics committee and regulatory bodies.
2. In the event that vulnerable subjects are enrolled in a clinical trial of a new chemical or molecular entity, an audio visual recording of the informed consent process must be maintained by the investigator. In the case of

HIV- and leprosy-related drugs, only the audio recording of the informed consent process needs to be maintained by the investigator (Government of India, Ministry of Health & Family Welfare [MOHFW], 2015, p. 3).

3. The investigator should explicitly inform the subject that the investigational product may fail to produce a desired therapeutic effect. The subject must also be informed that in placebo controlled trials, the placebo will have no therapeutic effect (MOHFW, 2015, p. 3).

CHAPTER REVIEW

In this chapter, we reviewed the FDA regulations and ICH GCPs related to clinical trial start-up. We looked at the various tasks required by the sponsor, the research site, and the investigator that need to be completed before clinical trial start. Multiple documents must be reviewed, signed, and returned to the sponsor by the research site. The protocol, informed consent, recruitment, and marketing materials must be approved by an appropriate IRB. The investigational product and clinical trial supplies must be shipped to the research site by the sponsor. Protocol-specific and EDC training must be completed by all users of the systems and be documented. All users of the EDC system must provide a written signature to the sponsor and be assigned login credentials and passwords for security. Budgets and contracts must be approved and signed. With all of this completed, once the protocol is approved the research site is ready to start the clinical trial. A research site initiation visit by the sponsor may be required prior to clinical trial start-up.

APPLY YOUR KNOWLEDGE

1. The clinical trial's principal investigator (PI) is currently the PI on 23 studies and practices full-time. He has signed the Form FDA 1572 and submitted it along with the other required forms to the sponsor. The research site and sponsor have completed all of the tasks required to start the clinical trial. The PI meets with the clinical trial coordinator a few days before the first subject is scheduled to be screened for the study. Because of his busy schedule, he is not able to devote a lot of time to the clinical trial. He instructs the coordinator to take the lead role in oversight of the clinical trial. Is this appropriate? Is it compliant with regulations and GCPs? How should the coordinator handle this situation?
2. You are preparing your research site for clinical trial start-up. You have received the clinical trial supplies and investigational product and have these stored and ready for the clinical trial. You have also received IRB approval for the research site's specific informed consent and the protocol and have submitted Form FDA 1572; investigator's CV and medical license, proof of insurance, and financial disclosures; and the investigator's delegation of authority to the sponsor. Your staff have reviewed the protocol and

have been trained on the protocol procedures. Is research site preparation complete or are there additional items/tasks required by the research site prior to starting the clinical trial?

REFERENCES

DiMasi, J. A., Grabowski, H. G., & Hansen, R. W. (2016, May). Innovation in the pharmaceutical industry: New estimates of R & D costs. *Journal of Health Economics*, *47*, 20–33. http://dx.doi.org/10.1016/j.jhealeco.2016.01.012

FDA IND Responsibilities of Sponsors and Investigators, 21 C.F.R. § 312.53 (2015).

Government of India, Ministry of Health & Family Welfare. (2015). *The gazette of India: Extraordinary, notification*. Retrieved from http://www.ferci.org/wp-content/uploads/2014/07/Gazette-Notification-31-July-2015-AV-consent.pdf

International Conference on Harmonisation. (1996). *Guideline for good clinical practice E6(R1)*. Retrieved from http://www.fda.gov/downloads/Drugs/GuidanceCompliance RegulatoryInformation/Guidances/ucm073122.pdf

U.S. Department of Health, Food and Drug Administration. (2013). *Statement of investigator*. Retrieved from http://www.fda.gov/downloads/AboutFDA/ReportsManualsForms/Forms/UCM074728.pdf

4 Enticement, Enrollment, and Engagement *The Informed Consent Process*

The informed consent process is one of the most important elements in the conduct of clinical trials. Yet, historically, the compromise of the ethical principles underlying informed consent has been the cause of some of the most troubling and ethically challenged research studies ever conducted. In this chapter, we will discuss the informed consent process from recruitment to screening, enrollment, and throughout the duration of the study.

CERTIFICATES OF CONFIDENTIALITY

Certificates of Confidentiality (CoCs) are not a standard element of informed consent but should be considered in preparation for studies that may include sensitive, confidential information from subjects. Although confidentiality is a key element of subject protection in clinical trials, there are studies that may place subjects at risk because compelled disclosure of sensitive personal information could have adverse consequences to their financial standing, employability, insurability, or reputation. Certificates of Confidentiality are issued by the National Institutes of Health (NIH) and allow investigators and/or others having access to research records to refuse disclosure of identifying information in any civil, criminal, administrative, legislative, or other proceeding, whether at the federal, state, or local level (U.S. NIH, 2016). Examples of potentially sensitive information could include the collection of genetic information, the collection of data on personal, illegal risk behaviors or the collection of information on sexual preferences or practices. Although sensitive information may be required for the conduct of a study, it could place the subject at risk should the information become public.

Investigators who are interested in obtaining a CoC for a study do not need to be affiliated with the NIH; any investigator conducting health-related research in which sensitive information will be collected from human research participants may apply for a CoC as long as the study has been approved by an institutional review board (IRB) that has an approved Federal-Wide Assurance or the approval of the Food and Drug Administration (FDA). Certificates are typically issued for a specific, well-defined research study (not a collection of studies) and remain in effect during the course of the study only. Instructions and applications for Certificates are submitted electronically through the NIH website.

INFORMED CONSENT PROCESS

The International Conference on Harmonisation (ICH) defines informed consent as "A process by which a subject voluntarily confirms his or her willingness to participate in a particular trial, after having been informed of all aspects of the trial that are relevant to the subject's decision to participate. Informed consent is documented by means of a written, signed, and dated informed consent form" (ICH, 1996, Section 1.28).

Informed consent is truly a *process* of sharing information and soliciting voluntary participation from a potential research subject before, during, and even after a study. The process is informed by the ethical principles of the Belmont Report (Appendix A), the Declaration of Helsinki (Appendix A), and the Nuremberg Code (Appendix A) and is codified through Federal Regulations in 21 C.F.R. 50.20-50. Each of these principles, codes, and regulations has the common intentions of voluntary consent (autonomy) of individuals, fair distribution of research risks and benefits, and protection of those persons with diminished autonomy.

The general requirements for informed consent in industry-sponsored studies can be found in 21 C.F.R. 50.20 and include the following four areas of interest:

- Except as provided in 50.23 and 50.24, no investigator may involve a human being as a subject in research covered by these regulations unless the investigator has obtained the legally effective informed consent of the subject or the subject's legally authorized representative (LAR).
- An investigator shall seek such consent only under circumstances that provide the prospective subject or the representative sufficient opportunity to consider whether or not to participate and that minimize the possibility of coercion or undue influence.
- The information that is given to the subject or the representative shall be in language understandable to the subject or the representative.
- No informed consent, whether oral or written, may include any exculpatory language through which the subject or the representative is made to waive or appear to waive any of the subject's legal rights, or releases or appears to release the investigator, the sponsor, the institution, or its agents from liability for negligence (FDA Protection of Human Subjects Informed Consent Requirements, 2015).

In summary, the general requirements for informed consent state that individuals should not be included in a study (or study procedures) without freely giving their explicit permission in a language that is understandable to them. Of course, there are exceptions to this statement, especially in the case where an investigator believes that an individual would benefit from the study but is unable to give consent. The "Exception from general requirements," 21 C.F.R. 50.23, requires that the investigator and a physician who is not otherwise participating in the clinical study certify in writing the following four conditions:

1. The subject is confronted by a life-threatening situation necessitating the use of the test article.

2. Informed consent cannot be obtained from the subject because of an inability to communicate with or obtain legally effective consent from the subject.
3. Time is not sufficient to obtain consent from the subject's legal representative.
4. There is no available alternative method of approved or generally recognized therapy that provides an equal or greater likelihood of saving the life of the subject (FDA Protection of Human Subjects Exception from General Requirements, 2015).

This regulation also addresses the waiver of consent in the use of investigational new drugs in connection with military operations. Under 10 U.S.C. 1107 (f), the President of the United States may waive the consent requirement for administration of an investigational new drug to a member of the armed forces in connection with the member's participation in a particular military operation (FDA Protection of Human Subjects Exception from General Requirements, 2015). The circumstances under which this can occur are detailed within 21 C.F.R. 50.23 but are related to feasibility, best interest of the military members, and national security.

One additional exception to informed consent is related to emergency research. In the case of life-threatening situations, treatment must be given quickly, and often times, an individual is unconscious, disoriented, or unable to make an informed choice. At the same time, an LAR might not be available. This type of consent waiver might relate to individuals who have suffered a head trauma, stroke, cardiac arrest, or other extreme health emergency.

Although there is a strong possibility that a clinical study in emergency research will include more than minimal risk to subjects, a waiver of consent may be applicable if there is a prospect of direct benefit to subjects. A waiver may be given if the following conditions are met:

- The subjects are in a life-threatening situation, available treatments are unproven or unsatisfactory, and the collection of valid scientific evidence is necessary to determine the safety and effectiveness of particular interventions.
- Obtaining informed consent is not feasible due to medical condition, an LAR is not available and there is no reasonable way to identify prospectively that the individual is likely to become eligible for the participation in the study.
- Participation offers the prospect of direct benefit to the subject.
- The study cannot practicably be carried out without the waiver.
- The proposed research plan defines the therapeutic window, and the investigator commits to trying to locate an LAR who can give consent within that window before proceeding to waive consent.
- The study and consent form (to be used when possible) has been approved by an IRB.
- Additional protections, including consultation with community representatives, public disclosure following completion of the study, establishment of an independent data monitoring committee, and a commitment to try to contact within a therapeutic window, the subject's family member who

is not an LAR to ask if there is an objection to the study (FDA Protection of Human Subjects Exception from Informed Consent Requirements for Emergency Research, 2015).

Clearly, exceptions to informed consent and emergency research require the investigator and his or her team to plan how they will operationalize a study that meets the parameters of these studies. This can be accomplished by having standard operating procedures in place and by including study stakeholders in the process planning from the very beginning. For example, if an investigator plans to conduct a myocardial infarction study through the emergency department of a cardiac hospital, she or he should include the physicians and staff who work in that department, community representatives (perhaps the local chapter of the American Heart Association), the IRB, nursing administration, hospital administration, and others who might be integral to the success of meeting the parameters defined by the study—this may also include personnel who have first contact with the subject, such as emergency transport (ambulance) personnel and admissions personnel.

RECRUITMENT AND SCREENING

The FDA considers direct recruitment/advertising for study subjects as the start of the informed consent and subject screening processes (U.S. FDA, 2016c, section A). It is during the recruitment process that potential subjects begin to understand the purpose, risks/benefits, time commitments, and location of the research. As such, recruitment methods and advertising should be included as part of an investigator's submission package to the IRB of record.

In its review of the submission packet, the IRB reviews the investigator's methods of recruiting study subjects, including any advertising, to ensure equitable selection of subjects per 21 C.F.R. 56.111(a)(3) and to ensure that the rights and welfare of subjects are protected, that is, Belmont principles of Justice and Respect for Persons, respectively. After reviewing the methods of recruitment to ensure that a particular population is not specifically targeted to accept greater risk or benefit than another, the IRB reviews advertisements for elements that might mislead subjects by implying a certainty of a favorable outcome/benefit or might present undue influence or coercion. For example, an advertisement for a "birth control" study might mislead potential subjects by implying that the study is offering "free" birth control when the investigational drug's effectiveness as a birth control method is still being evaluated during the study. Another study advertisement might be considered coercive in nature if subject compensation and distribution unduly influence a destitute person to accept a high level of risk to participate in a study.

SCREENING

During the study feasibility process, investigators determine if they have access to a population of individuals who might qualify for the study under negotiation. Research sites that conduct similar disease-specific studies or have access to a patient population that has given permission to be contacted with potential research studies have an advantage over research sites that have not identified a population before a study starts. However, there are no guarantees that all the individuals in a specific population will be eligible to enroll in a specific study. Each study has its own eligibility criteria that define the inclusion criteria as well as the exclusion criteria for participation. Before study volunteers can be enrolled in a study, they must be "screened" to determine whether they meet the eligibility criteria.

An investigator can discuss the availability of a study with a potential subject without first obtaining consent, but informed consent must be obtained before any clinical procedures are conducted solely to determine eligibility for research (U.S. FDA, 2016d, para. 1). As an example, a gastroenterologist, who is also an investigator for a Crohn's disease study, may discuss the availability of a study with one of his patients with the disease. However, if eligibility criteria include withdrawal of all medication before the start of the study, informed consent must be obtained before the medication is withdrawn. In some cases, study eligibility may require a study exam, such as a radiograph, within a specific timeframe. If a radiograph was completed but needs to be repeated to meet the timeframe for eligibility, informed consent must be obtained before the radiograph is repeated.

Not every volunteer enrolls in a study, either because she or he does not meet the eligibility criteria or because she or he declines to participate. Sponsors often request site personnel to maintain a screening log that tracks the status of potential subjects screened, as well as the number of "screening failures." A screening log should not include identifiable information regarding the volunteer (to be Health Insurance Portability and Accountability Act [HIPAA] compliant) but should indicate whether "screened" individuals were eligible and were enrolled, ineligible (and why), or declined to participate (and why). Typically, the sponsor and/or the protocol defines how a screening failure should be determined, but in most cases, the term *screening failure* refers to the number of volunteers who signed an informed consent and may have had screening procedures but did not have any other study-related procedures.

A screening log is a best practice in study documentation and can help define metrics for each study. The metrics, in turn, can assist research site management to make better decisions on recruitment strategies, sponsor relationships, and budgeting.

INFORMED CONSENT DOCUMENT AND DOCUMENTATION

A consent form is a tool used in the informed consent process to document a subject's consent to participate in a research study. The form has been approved by an IRB and it contains at least eight "basic" elements and potentially six additional elements that should be included when appropriate and/or when required by the study sponsor (as found in 21 C.F.R. 50.25). Because many studies are conducted globally, most sponsors require the consent form to include the required elements found in Section 4.8.10 of ICH E6 (Appendix A) to ensure that consent forms are similar across all research sites regardless of site location.

Although the FDA does not require the investigator to personally conduct the consent interview during the consent process, she or he remains ultimately responsible, even when delegating the task to another person who is knowledgeable about the research. The consent form should be used to help the investigator explain the study and to initiate an exchange of information. It is the investigator's responsibility to verbally explain the study and emphasize that the study is research (as opposed to standard care). Then she or he should discuss the purpose of the research, a description of risks and benefits, alternatives to the study (treatments or other options), study procedures, description of confidentiality, compensation, and the other basic and required elements found within the consent form. During the explanation, the investigator should not be hurried, should pause to allow questions and/or comments, and take time to explain items. Oftentimes, investigators use open-ended questions to ensure that the potential subject understands key aspects during the explanation. For example, "Just so that I am sure you understand the alternatives to this study, could you explain, in your own words, what other treatment options you have if you don't want to participate in this study?" Most open-ended questions begin with terms such as "what," "how many," and "could."

Once the investigator has verbally explained the study, the investigator allows potential subjects sufficient time to read, review, discuss, and confer with family and friends to consider whether they want to participate in the study. Obviously, "sufficient time" is a relative term. Reasonable timeframes for decisions can range from hours to days or weeks, depending on how the potential subject evaluates the risks, benefits, or alternative treatment options.

After allowing time for potential subjects to make a decision, the investigator who is named in the study/consent form should meet with potential subjects to answer any further questions and ensure that they comprehend the study. Upon receiving their voluntary consent, the study subjects sign and date the written consent as documentation of their permission to participate in the study.

Consent Documents

Informed consent is documented by the use of a written consent form that must be signed by the subject or by the subject's LAR, as determined by state law. The consent, as mentioned previously, should include the basic and applicable additional elements of informed consent. Regulations require that a copy of the informed consent form (ICF) be given to the subject; however, federal regulations do not require that it be a photocopy of the signed consent form. Conversely, ICH guidelines, HIPAA authorization criteria, some IRBs, and some states require that subjects receive a copy of the "signed" consent form. As such, it is good practice to photocopy the signed consent form and give a copy to the subject.

In specially approved situations, a "short form" may be used during the informed consent process if an investigator unexpectedly encounters a nonEnglish-speaking subject. If an investigator enrolls a subject without having/using an IRB-approved translated consent form, she or he may use a short form, written in a language understood by the subject, to document that the elements of informed consent found in 21 C.F.R. 50.25 were orally presented in a language understood by the subject (U.S. FDA, 2016a).

The process of administering the short form is different from that of the full-consent process. The content of the short form and a summary of what is to be said to the subject must be approved by an IRB (prior to subject delivery), and there must be a witness present during the oral presentation who understands and speaks both English and the subject's native language. The witness and the interpreter can be one and the same person. The witness signs both the short form and the summary document attesting that the information in the summary document was presented to the subject in a language understandable to the subject, that the subject's questions were presented and answered in a language understandable to the subject, and that the subject was asked in a language understandable to him or her if he or she understood the information and that he or she responded to affirmatively. The subject signs the short form only, and the person obtaining the consent must sign the summary only. A copy of the summary and the short form must be given to the subject (FDA Protection of Human Subjects Documentation of Informed Consent, 2015).

Most IRB applications ask whether the investigator plans to recruit nonEnglish-speaking populations. If so, the IRB requires investigators to have translated consent forms, interpreters, and plans in place to ensure that subjects are consented in a language understandable to them.

Recruiting nonEnglish-speaking subjects requires diligence on the part of the investigator and his or her team. Standard operating procedures should be in place to ensure that processes are not overlooked and that procedures are efficient and well planned. Additionally, if the study team intends on creating and maintaining a collaborative relationship with a particular cultural group, it is important to include members from that group in the research team, as employees, interpreters, and/or liaisons with the community.

Signatures

To document a subject's consent, she or he signs and dates the consent form. The investigator or his or her delegate who led the consent interview should also sign and date the forms. Once signed, the subject should be given a copy of the consent form and the original consent form should be kept in the research site study file.

The date and time of the consent should be documented in the medical records or notes. This is particularly important to provide evidence that consent was obtained prior to initiation of any study procedures, including screening examinations.

VULNERABLE SUBJECTS

Individuals are considered vulnerable when they do not have full autonomy in making informed decisions that may impact their own interests and safety. Vulnerable subjects are individuals who cannot give verbal or signed consent because of their physical/mental status or because of special circumstances. Young children, unconscious individuals, or cognitively impaired patients are considered vulnerable, and other individuals who are pregnant or imprisoned may be considered vulnerable because of their current situation. The Belmont Report recognizes that vulnerabilities arise from social, racial, sexual, and cultural biases in society and that researchers must be aware of these situations to ensure that certain populations are not chosen for research studies as a convenience only.

Children

Children are addressed as vulnerable subjects in both FDA regulations (21 C.F.R. 50, Subpart D) and Department of Health and Human Services (HHS) regulations (45 C.F.R. 46, Subpart D). However, it has only been recently that the FDA has encouraged the inclusion of children in clinical trials to ensure drug safety and effectiveness in children. Previously, only about 20% of drugs approved by the FDA were labeled for pediatric use, necessitating off-label use of adult-tested drugs in children (U.S. FDA, 2011, para. 2).

The FDA defers to individual states for legal age and age-related regulations regarding child inclusion and minors in research. The definition of "child" varies state to state, but typically, children are individuals who have not reached the age of 18. Because some states acknowledge special circumstances for some children, it is important that the investigator and team are aware of state laws when recruiting children as subjects in a study.

To be enrolled in a clinical trial, the parent(s) or guardian(s) of the child must give permission for the child to participate. Additionally, when appropriate, the child should give assent to participate. " 'Assent' means a child's affirmative agreement to take part in a clinical investigation, not just the failure to object" (U.S. FDA, 2016b, part F). In other words, the child is making a decision to participate in a study.

Although the investigator is ultimately responsible for the conduct of the study, the FDA places enormous responsibilities on the IRB approving the study to make determinations regarding consent and assent, as well as other decisions that may impact the safety and autonomy of the child. The IRB ensures that adequate provisions are made for soliciting assent of children, considers whether to require the assent of children in addition to their parents, and determines how assent is documented.

Clearly, enrolling children in clinical trials requires the research site to have processes and procedures in place to accommodate the requirements of the regulations as well as the IRB. Additionally, the investigator and team must have a strong relationship with the IRB to ensure that provisions that are placed on the trial are appropriate and will protect the child's safety.

Other Vulnerable Populations

The FDA recognizes that there are other vulnerable populations besides children and asks the approving IRB to ensure that there is equitable treatment of these populations. Title 21 notes that "the IRB should take into account the purposes of the research and the setting in which the research will be conducted and should be particularly cognizant of the special problems of research involving vulnerable populations, such as children, prisoners, pregnant women, handicapped, or mentally disabled (cognitively impaired) persons, or economically or educationally disadvantaged persons" (FDA Institutional Review Boards Criteria for IRB Approval of Research, 2015).

Cognitively Impaired Individuals

Cognitively impaired individuals do not possess full capacity for judgment and/or reasoning due to a temporary acute condition such as a seizure or as a result of a permanent condition such as Alzheimer's disease. Conducting clinical research to advance knowledge on conditions that result in cognitive impairment is vital, yet the conditions themselves may compromise a potential research subject's ability to give legally effective informed consent to participate in the research. Individuals who are cognitively impaired may not have the capacity to make meaningful decisions in regard to study participation. Because of this unique circumstance, the IRB and principal investigator must consider and/or implement special protections, such as involving someone independent of the study, for example, an unaffiliated clinician, to serve as a monitor of the consent process, to safeguard this population (U.S. NIH, 2009).

Decision-making capacity is specific to particular situations and particular protocols. The setting and situation in which consent is sought should be conducive to subject appreciation and comprehension. For example, an individual may have the capacity to consent to a protocol that is easy to understand and is of low risk but may not have the capacity to understand a complex protocol that includes high risk. It is the principal investigator's responsibility to work with the IRB to develop a plan for assessing potential subjects. Although there are no generally accepted criteria for determining competence to consent to research, many investigators have developed various approaches to assess a potential subject's ability to give informed consent. Some investigators use a screening standard mental status examination such as the MINI-Mental Status Exam or the Alzheimer's Disease Assessment Scale-Cognitive Portion assessment to assess decision-making capacity. However, these may not be definitive and the potential subject might require further assessment by a physician/psychologist.

Enrolling individuals with impaired consent capacity may necessitate the assistance and/or involvement of an LAR, who acts on behalf of a prospective subject. This is a specific consideration that should be addressed on the recruitment plan before the IRB approval process to ensure communication and continued informed consent of subjects through their LAR throughout the study. If it is likely that a subject's consent capacity will deteriorate during a study, it is prudent to involve an LAR from the beginning of the study to facilitate a transition to LAR consent and decision making later in the study.

Safeguards, processes, and procedures vary depending on subject population, type of cognitive impairment, and complexity of study. It is imperative that an investigative team and IRB weigh the impact of risks and benefits on potential subjects and employ safeguards to ensure the safety and continued consent in this vulnerable population.

Pregnant Women, Human Fetuses/Neonates, and Prisoners

The HHS specifically mentions two populations that require special protections. Title 45, Part 46, Subpart B, addresses "Additional Protections for Pregnant Women, Human Fetuses and Neonates Involved in Research," and Title 45, Part 46, Subpart C, addresses "Additional Protections Pertaining to Biomedical and Behavioral Research Involving Prisoners as Subjects."

There are specific conditions under which pregnant women or fetuses can participate in research, according to 45 C.F.R. 46.204, including the following (see 45 C.F.R. Part 46, Subpart B for full requirements):

- Preclinical studies, including studies on pregnant animals, and clinical studies, including studies on nonpregnant women, have been conducted and provide data for assessing potential risks to pregnant women and fetuses;
- The risk to the fetus is caused solely by interventions or procedures that hold out the prospect of direct benefit for the woman or the fetus; or, if there is no such prospect of benefit, the risk to the fetus is not greater than minimal risk to the fetus;
- Any risk is the least possible for achieving the objectives of the research;
- If the research holds out the prospect of direct benefit solely to the fetus, then the consent of the pregnant woman and the father is obtained in accord with the informed consent provisions, except that the father's consent need not be obtained if he is unable to consent because of unavailability, incompetence, or temporary incapacity or the pregnancy resulted from rape or incest;
- No inducements, monetary or otherwise, will be offered to terminate a pregnancy;
- Individuals engaged in the research will have no part in any decisions as to the timing, method, or procedures used to terminate a pregnancy or determining the viability of a neonate (Research Involving Pregnant Women or Fetuses, 2010).

Prisoners who participate in clinical research are susceptible to coercion and influence because of the nature of their confinement, which includes limited freedom to make decisions. Any research that is conducted in a prison must be approved by an IRB that does not have any association with the prison where the study will be conducted. Additionally, the IRB must include at least one member who is a prisoner or a prisoner representative. Other requirements for prisoner protection include the following (see 45 C.F.R. Part 46, Subpart C for full requirements):

- Any possible advantages to the prisoner through his or her participation in the research are not of such a magnitude that his or her ability to weigh the risks of the research against the value of such advantages in the limited choice environment of the prison is impaired;
- The risks involved in the research are commensurate with risks that would be accepted by nonprisoner volunteers;
- Procedures for subject selection within the prison are fair to all prisoners and immune from arbitrary intervention by prison authorities or prisoners;
- The information is presented in language which is understandable to the subject population;
- Adequate assurance exists that parole boards will not take into account a prisoner's participation in the research in making decisions regarding parole;
- Where the Board finds there may be a need for follow-up examination or care of participants after the end of their participation, adequate provision has been made for such examination or care (Permitted Research Involving Prisoners, 2009).

Although the federal regulations address children, women, and prisoners as vulnerable populations, there are other populations, individuals, and groups that may be at risk of being vulnerable during specific timeframes. Section 1.61 of the ICH E6 Guidelines (Appendix A5) notes the following:

> Individuals whose willingness to volunteer in a clinical trial may be unduly influenced by the expectation, whether justified or not, of benefits associated with participation, or of a retaliatory response from senior members of a hierarchy in case of refusal to participate. Examples are members of a group with hierarchical structure, such as medical, pharmacy, dental, and nursing students, subordinate hospital and laboratory personnel, employees of the pharmaceutical industry, members of the armed forces, and persons kept in detention. Other vulnerable subjects include patients with incurable diseases, persons in nursing homes, unemployed or impoverished persons, patients in emergency situations, ethnic minority groups, homeless persons, nomads, refugees, minors, and those incapable of giving consent. (ICH, 1996, Section 1.61)

Clearly, undue influence to enroll in a research study may arise from many different situations and circumstances. As such, the investigator should be sensitive to the study population and potential vulnerabilities.

IRB Recruitment Plan of Vulnerable Populations

The IRB in charge of reviewing a study is charged with reviewing an investigator's IRB application to determine if the study might include a vulnerable group. As an example, the IRB might ask an investigator whether he intends to recruit volunteers from his investigational team. The IRB considers this a hierarchical situation because the investigator might intend on asking employees to participate as part of their performance evaluation creating a situation where the investigator has undue influence on a subject.

To align with the ethical foundations and regulations regarding vulnerable populations, it is very important that the investigator has a recruitment plan developed by the time the IRB application is completed. If a vulnerable population is the target of the research, the investigator should have appropriate reasons why the population is being targeted. Additionally, she or he should have an outline of how the recruitment will take place, detailing the circumstances, personnel involved, and any scripts that will be used. As mentioned previously, research sites with standard operating procedures, processes, forms, and checklists will be more likely to be compliant with the regulations and ethical considerations for all aspects of a study.

HEALTH INSURANCE PORTABILITY AND ACCOUNTABILITY ACT

In March 2016, a clinical research facility agreed to pay the U.S. HHS Office for Civil Rights (OCR) a $3.8 million settlement for a HIPAA violation that included a security breach, making this the first research site settlement for a HIPAA security breach in the OCR's history. The OCR director, Jocelyn Samuels, noted, "Research institutions subject to HIPAA must be held to the same compliance standards as all other HIPAA-covered entities" (U.S. HHS, 2016b, para. 4). The HIPAA includes two sections, privacy and security rules.

Privacy Rule

The goal of the HIPAA Privacy Rule is to ensure that individuals' health information is properly protected while allowing the flow of health information needed to provide health care and to protect the public's health and well-being (U.S. HHS, 2003, para. 2). The HIPAA rules apply to covered entities and business associates. A covered entity is a health care provider, a health plan, or a health care clearinghouse. Typically, research sites fall under the category of health care provider if they are part of a doctor's office, clinic, nursing home, or hospital and "they transmit information in an electronic form in connection with a transaction for which HHS has adopted a standard" (U.S. HHS, 2016a). Transactions typically concern billing and payment for services or insurance coverage.

The impact on research is to require specific consent from subjects regarding the use of their protected health information (PHI) in clinical trial reporting. For many organizations, HIPAA consent is often situated in the same office as the IRB. There are five different ways in which researchers can access PHI for research, including the following:

- Authorization form—A research subject signs an authorization form to allow his or her PHI to be used for research purposes. Authorization is required in addition to obtaining informed consent to participate in research. It focuses on privacy risks and states how and why the PHI will be used and to whom PHI will be disclosed for research.
- HIPAA Waiver or Alteration of Authorization—A waiver can be requested when a researcher is unable to use deidentified health information and the research cannot practicably be conducted. This requires the researcher to justify how the research is to be conducted and why it cannot practicably be conducted without a waiver or without access and use of PHI.
- Limited data set with a data use agreement—A researcher may use a limited data set after establishing a data use agreement with the holder of the PHI. A limited data set refers to PHI that excludes 16 categories of direct identifiers, including names, addresses, telephone numbers, fax numbers, email addresses, social security numbers, medical record numbers, health plan beneficiary numbers, account numbers, certificate/license numbers, vehicle identification numbers, device identifiers and serial numbers, website URLs, IP addresses, biometric identifiers, and full-face photographic images.
- Activities preparatory to research—A researcher may use PHI for activities involved in preparing for research without an authorization or waiver upon agreement (orally or in writing) that the use is solely in preparation for research; the PHI will not be removed from the covered entity in the course of review; and the PHI is necessary for the research.
- Decedents' PHI—Research that uses/discloses PHI is covered by the HIPAA Privacy Rule and must comply with HIPAA regulations. Authorization from personal representatives, waivers, and/or data use agreements are not required by HIPAA to use decedent PHI.

Security Rule

The HIPAA Security Rule goal is to protect the privacy of health information while allowing covered entities to adopt new technology to improve the quality and efficiency of patient care. Covered entities must

- Ensure the confidentiality, integrity, and availability of all electronic PHI (ePHI) they create, receive, maintain or transmit;
- Identify and protect against reasonably anticipated threats to the security or integrity of the information;
- Protect against reasonably anticipated, impermissible uses or disclosures; and
- Ensure compliance by their workforce (U.S. HHS, 2016c).

In summary, it is the research site (covered entity) that is responsible for any inadvertent release of PHI that may occur while handling ePHI. In the example cited at the beginning of this section, the security breach occurred when an employee borrowed a laptop from work that contained records from 13,000 patients. The laptop was subsequently stolen from a vehicle he was using.

Obviously, laptops and other mobile devices are lost and stolen on a regular basis. However, the OCR expects that entities use best practices in personnel policies, standard operating procedures, employees' use of ePHI or devices that contain ePHI, security software, access policies, offsite computer network access, encryption, workforce training, and more. This can be very expensive, especially if a covered entity hires information security consultants to assist with HIPAA compliance. However, should there be a disclosure or other type of breach of ePHI, the OCR determines whether an entity made attempts to be compliant and the extent of their success when the Department determines penalties.

ONGOING CONSENT

As discussed throughout this chapter, informed consent is a process of sharing information and affirming the subject's continued consent to participate in a study. Ethical conduct of research requires that subjects are informed of new findings, new risks/benefits, or other information that may influence their continued participation in the study and/or follow-up. Significant new findings may include an unexpected adverse event or an adverse event that occurs at greater frequency or severity than was previously stated in the consent form (U.S. FDA, 2016b, section III).

There may be times throughout a study that additional risks and/or benefits are identified that may differ significantly from those identified in the original consent form. In these cases, a sponsor will most likely amend the protocol and/or consent form to identify these risks or other considerations. Reconsent of the subjects who are already enrolled may be appropriate and/or required. Reconsent may also be appropriate when children reach adulthood so that research participation reflects their choices, rather than those of their parents or guardians. These decisions should be made in concert with the IRB.

Reconsent is often conducted in the same manner as the original consent process, depending on the circumstances of the study and the changes involved. The investigator explains the need for reconsent and gives subjects the opportunity to ask questions. Should the subjects wish to continue in the study, they sign a modified consent form that has been approved by the IRB, affirming their willingness to continue to participate in the clinical trial.

A VIEW FROM INDIA

The informed consent process in India has similar ethical and scientific guidelines as those in the United States. However, there are aspects of Indian laws and guidelines that are specific to the social and cultural environment in India.

Certificates of Confidentiality

In India, CoCs are not an option for researchers. Rather, the duty to safeguard the identification of clinical subjects and the confidentiality of a clinical trial's data rest with the sponsors, the ethics committees, and the principal investigators who conduct the clinical trial. However, data of individual subjects may be disclosed under the following circumstances:

1. By orders of a judge, only in an applicable court of law;
2. A threat to the subject's life;
3. To communicate to the drugs regulatory authority, in case of a serious adverse event; and
4. To communicate to the health authorities, in case of risk to public health, precedence over personal right to privacy may be taken (Indian Council of Medical Research New Delhi, 2006, p. 29).

Informed Consent Process

In addition to the U.S. guidelines and regulations in this chapter, Indian guidelines recommend that

1. A separate consent must be obtained for publication of data (appropriately camouflaged photographs or pedigree charts) and
2. This consent should be obtained on a different occasion and not as part of the consent where the subject consents to participate in the clinical study.
3. In India, if genetic or HIV testing is part of the trial, counseling for consent for testing must be given as per national guidelines (Ministry of Health & Family Welfare [MOHFW], 2013).

Informed Consent Document and Documentation

In India, the informed consent document (ICD) contains the patient information sheet (PIS) and the ICF. The PIS contains at least 14 essential elements and potentially six additional elements that should be included when appropriate and/or when required by the study sponsor. A checklist of the essential and additional elements to be included in the study subject's PIS as well as a format for the ICF is given in Appendix V of Schedule Y as amended in 2013.

As mentioned in the previous chapter, it is required that an ICD is translated into the language spoken by the local population being recruited by the research site. Back-translation to English is recommended and may be required by some IRB/independent ethics committee (IEC) and regulatory bodies. The person who translates the ICD and the person who back-translates the ICD should be different. Translation and back-translation of ICDs must be approved by the IRB/IEC. Additionally, prior to administration of an ICF, there must be adequate pretesting of the ICD to determine that it is comprehensible to individuals who will be recruited for the research trial.

When subjects sign a consent form in their known language, the person administering the consent must also sign in that language. In other words, both signatures must be in the same language; for example, a Hindi-language ICF requires that signatures are also in Hindi.

As is true in the United States, voluntary participation of a subject means that refusal to participate in a research trial or withdrawal from a research trial will not influence the patient/clinician relationship; or the benefits, which the subject is otherwise entitled to.

Vulnerable Subjects

The ethics committee is responsible to protect the rights, safety, and well-being of all vulnerable subjects participating in a research trial. The following categories of subjects can be considered as vulnerable: patients with incurable diseases, patients in an emergency situation, members of a group with hierarchical structure (for example, prisoners, armed forces personnel, staff and students of medical, nursing and pharmacy academic institutions), ethnic minority groups, unemployed or impoverished persons, homeless persons, nomads, refugees, minors, or others incapable of personally giving consent (Government of India, MOHFW, Central Drugs Standard Control Organization, 2004).

Children

The investigator must be responsible and take care of the following considerations before undertaking clinical trials in children:

1. Children cannot be involved in a research study that can be carried out equitably with adults,

2. Clinical evaluation of a new drug study in children should be carried out only after phase III clinical trials in adults are completed. However, if the drug has a therapeutic value in a primary disease of children, it can be studied earlier;
3. A proxy consent for each child is given by a parent or a legal guardian and either an oral or a written assent of the child is obtained based on his or her age appropriate capabilities; and
4. The risks and benefits of the intervention have been carefully assessed (Indian Council of Medical Research New Delhi, 2006, p. 28).

Pregnant Women, Human Fetuses/Neonates, and Prisoners

Pregnant or lactating women can participate in clinical trials that include investigations, drugs, vaccines, or other agents that promise therapeutic or preventive benefits; for example, to test the efficacy and safety of a drug for reducing perinatal transmission of HIV infection from mother to child (Indian Council of Medical Research New Delhi, 2006, p. 27). Pregnant women can participate in research related to prenatal diagnostic techniques when such research is limited to detecting fetal abnormalities or genetic disorders as per the Prenatal Diagnostic Techniques (Regulation and Prevention of Misuse) Act, GOI, 1994, and not for sex determination of the fetus (Indian Council of Medical Research New Delhi, 2006, p. 28). Pregnant women who desire to undergo medical termination of pregnancy are eligible to participate in research related to termination of pregnancy as per The Medical Termination of Pregnancy Act, GOI, 1971 (Indian Council of Medical Research New Delhi, 2006, p. 27).

Ongoing Consent

Subjects whose consent is waived in emergency settings should be consented during the recovery process when they are mentally competent to understand the research trial (Indian Council of Medical Research New Delhi, 2006, p. 22).

CHAPTER REVIEW

In this chapter, we examined the informed consent process beginning with recruitment, then we discussed the consent interview, consent documentation, vulnerable populations, HIPAA, and ongoing consent.

APPLY YOUR KNOWLEDGE

1. Mr. Marsh is very excited about joining an osteoarthritis research study for his knee because his current treatment does not give him the quality of life he wants. He researches the ClinicalTrials.gov website to find a study that he thinks will offer him some hope. He carefully reads the inclusion/exclusion criteria and discovers that to be eligible, he must not have taken any medication for his osteoarthritis within the last 2 weeks. Mr. Marsh

immediately stops his medication (without asking his doctor) and calls the research site to make an appointment in 2 weeks to learn more about the study. Are there any concerns in regard to this situation? What are the next steps?

2. Dr. Andrews is very passionate about a new treatment regime for asthma and has developed an outline for an investigator-initiated study. She currently collaborates with a clinic that treats a high number of asthma sufferers and would like to review the clinic's records to determine if her study outline is appropriate for this population and whether there is a population that might participate in the study. What steps does she need take to be able to review the clinic's database of PHI?

REFERENCES

FDA Institutional Review Boards Criteria for IRB Approval of Research, 21 C.F.R. § 56.111 (2015).

FDA Protection of Human Subjects Documentation of Informed Consent, 21 C.F.R. § 50.27 (2015).

FDA Protection of Human Subjects Exception from General Requirements, 21 C.F.R. § 50.23 (2015).

FDA Protection of Human Subjects Exception from Informed Consent Requirements for Emergency Research, 21 C.F.R. § 50.24 (2015).

FDA Protection of Human Subjects Informed Consent Requirements, 21 C.F.R. § 50.20 (2015).

Government of India, Ministry of Health & Family Welfare. (2013). *The Drugs and Cosmetics Act, 1940 and rules, 1945 Schedule Y*. Retrieved from http://www.mohfw.nic.in/WriteReadData/l892s/43503435431421382269.pdf

Government of India, Ministry of Health & Family Welfare. (2015). *The Gazette of India: Extraordinary, notification*. Retrieved from http://www.ferci.org/wp-content/uploads/2014/07/Gazette-Notification-31-July-2015-AV-consent.pdf

Government of India, Ministry of Health & Family Welfare, Central Drugs Standard Control Organization. (2004). *Good clinical practice for clinical research in India*. Retrieved from http://cdsco.nic.in/html/D&C_Rules_Schedule_Y.pdf

Indian Council of Medical Research New Delhi. (2006). *Ethical guidelines for biomedical research on human participants*. Retrieved from http://www.icmr.nic.in/ethical_guidelines.pdf

International Conference on Harmonisation. (1996). *Guideline for good clinical practice E6(R1)*. Retrieved from http://www.fda.gov/downloads/Drugs/GuidanceComplianceRegulatoryInformation/Guidances/ucm073122.pdf

Permitted Research Involving Prisoners, 45 C.F.R. § 306 (2009).

Research Involving Pregnant Women or Fetuses, 45 C.F.R. § 46.204 (2010).

U.S. Department of Health and Human Services. (2003). Summary of the HIPAA privacy rule. Retrieved from http://www.hhs.gov/hipaa/for-professionals/privacy/laws-regulations/index.html

U.S. Department of Health and Human Services. (2016a). Covered entities and business associates. Retrieved from http://www.hhs.gov/hipaa/for-professionals/covered-entities/index.html

U.S. Department of Health and Human Services. (2016b). Improper disclosure of research participants' protected health information results in $3.9 million HIPAA settlement. Retrieved from http://www.hhs.gov/about/news/2016/03/17/improper-disclosure-research-participants-protected-health-information-results-in-hipaa-settlement.html#

U.S. Department of Health and Human Services. (2016c). Summary of the HIPAA security rule. Retrieved from http://www.hhs.gov/hipaa/for-professionals/security/laws-regulations/index.html

U.S. Food and Drug Administration. (2011). Drug research and children. Retrieved from http://www.fda.gov/drugs/resourcesforyou/consumers/ucm143565

U.S. Food and Drug Administration. (2016a). A guide to informed consent—Information sheet: Guidance for institutional review boards and clinical investigators. Retrieved from http://www.fda.gov/regulatoryinformation/guidances/ucm126431

U.S. Food and Drug Administration. (2016b). Informed consent information sheet: Guidance for IRBs, clinical investigators, and sponsors. Retrieved from http://www.fda.gov/regulatoryinformation/Guidances/ucm404975

U.S. Food and Drug Administration. (2016c). Recruiting study subjects—information sheet: Guidance for institutional review boards and clinical investigators. Retrieved from http://www.fda.gov/regulatoryinformation/guidances/ucm126428

U.S. Food and Drug Administration. (2016d). Screening tests prior to study enrollment—Information sheet: Guidance for institutional review boards and clinical investigators. Retrieved from http://www.fda.gov/regulatoryinformation/guidances/ucm126430

U.S. National Institutes of Health. (2009). Research involving individuals with questionable capacity to consent: Points to consider. Retrieved from http://grants.nih.gov/grants/policy/questionablecapacity.htm

U.S. National Institutes of Health. (2016). General information on certificates and the protections provided. Retrieved from https://humansubjects.nih.gov/coc/faqs#general-info

5 From Enrollment to Final Visit

Data from clinical trials are used by the Food and Drug Administration (FDA) to determine approval of an investigational drug or device for market. It is paramount to safely marketing the drug that all data collected are accurate and complete. The FDA has regulations that clearly define how data should be collected, documented, and managed. Good clinical practices (GCPs) also provide guidance on conducting study visits and collecting data.

This chapter will review subject study visits, what occurs at study visits, the regulations defining the conduct of clinical studies, and collecting complete and accurate data. Complete and accurate data documentation includes data collection, data entry, data queries, data correction, and proper management and storage of study data. Not only is subject data protected by FDA regulations, it also falls under the purview of the Department of Health and Human Services, detailed in the Health Insurance Portability and Accountability Act (HIPAA).

PRESCREENING

During prescreening, the research team may conduct prescreening telephone interviews as an opportunity to save time and the financial costs of bringing potential subjects on site. During this phone call, the interviewer asks if the individual is willing to answer a few questions to determine if the study is appropriate for him or her. The interview includes a set of questions, based on study eligibility criteria, to determine if the individual meets the general eligibility requirements and does not have or has not been diagnosed with diseases or conditions that will disqualify him or her from the study. If the individual qualifies after the phone screen and is still interested in the study, she or he is invited to the research site for a screening visit.

CLINICAL TRIAL VISITS

The study protocol defines the services and procedures that occur at each study visit. During a study visit, the data required to prove safety and efficacy of the investigational product are collected. Each visit, procedure, and service is defined in detail in the study protocol, including possible side effects and adverse events (AEs) to the various procedures. The schedule of events (SE) found in the protocol outlines the number of visits and the procedures that occur at each visit.

The number of visits required and the procedures required at each visit depend on the type of study. Phase 1 studies are generally short and may only last 1 month. Phase 2 and 3 studies can run for years depending on the investigational product being tested. Procedures are based on the collection of data needed to prove the safety and efficacy of the investigational product. Many studies also include substudies, which may be used to collect additional information. Often, subjects are asked to agree to the collection of additional blood or other biologic samples that may be used in precision medicine studies or other studies using molecular biology. If a subject decides not to participate in a substudy, it may or may not impact the subject's participation in the main study—the protocol and informed consent form need to be very clear on whether a substudy is required as part of the primary study.

To determine how many visits are included in a particular study, the study protocol must be reviewed. The SE, a table of study visits and the procedures to be performed at each visit, is a good place to start. The SE should be combined with the rest of the protocol to determine all visits and procedures. For example, additional follow-up visits may be required as a result of an AE, such as an abnormal lab result. This information would not be in the SE but found elsewhere in the protocol. Table 5.1 provides an example of an SE that might be found in a protocol.

TABLE 5.1
Schedule of Events (Example)

Procedure	Screening Visit	Visit 1 Day 1	Visit 2 Week 2	Visit 3 Week 4	Visit 4 Week 8	Visit 5 Week 12
Consent	X					
Height	X					
Weight	X	X	X	X	X	X
Vitals	X	X	X	X	X	X
Physical	X					X
Blood draw	X	X		X	X	X
Urine Pg test	X	X	X	X	X	X
EKG	X	X				X
ConMeds	X	X	X	X	X	X
Adverse events		X	X	X	X	X
QOL survey	X	X		X		X
Randomization		X				
Distribute drug		X	X	X	X	

Screening Visit

The first study visit is typically a screening visit to determine if an individual meets the eligibility criteria. Informed consent must be obtained from the potential subject before any procedure can be conducted. The informed consent process involves much more than handing the informed consent form to subjects, instructing them to read it, and returning to answer any questions they may have, as was discussed in Chapter 4.

Once the signed and dated informed consent is obtained, the research team can proceed with the screening process. Results from the procedures and tests performed at the screening visit determine whether an individual is eligible for a study.

Each protocol includes subject eligibility criteria—what qualifies a subject to be included in the study and what excludes a potential subject from participating. These criteria are based on the disease or condition being studied, the type of investigational drug, and other factors identified during preclinical studies. The potential subject must meet all of the inclusion criteria. On the other hand, the subject cannot exhibit any of the exclusion criteria. Below are examples of inclusion and exclusion criteria. Generally, the exclusion criteria are a much longer list than the inclusion criteria.

Inclusion criteria

- Male or female
- 40 to 65 years of age
- Otherwise healthy
- Diagnosed with study specific disease or condition within the past 6 months
- Able to read, understand and be willing to sign the informed consent

Exclusion criteria

- Under 40 or older than 65 years of age when consented
- No diagnosis of study specific disease or condition
- Currently pregnant or plans to get pregnant within the next 12 months or is breastfeeding
- Not willing to use two forms of birth control
- Smokes more than 10 cigarettes per day
- Drinks more than 14 alcoholic beverages per week

The sponsor generally provides an eligibility checklist to the research site for use in determining whether individuals meet the criteria. It is the investigator's responsibility to review lab results and other data using the checklist and protocol to make the eligibility determination. Individuals who are eligible are scheduled for the first study visit.

Scheduled Visits

The procedures and tests listed for each visit determine the skills and/or licensure of members of the research team needed to properly conduct the visit and the amount of time required for each visit. This is defined in the protocol. For example, a protocol may require that the study physicals be conducted by a nurse practitioner rather than a registered nurse.

Depending on the type of research site, laboratory tests may be conducted onsite or outsourced to an independent laboratory. In some cases, sponsors employ a central laboratory where all study specimens are sent. The central lab analyzes the specimens according to the study protocol and sends the results to the research site for review and inclusion in the subject's record.

The majority of investigational drug studies are double-blind, randomized studies. In double-blind studies, neither the investigator nor the subject knows whether the subject is randomized to the control group or the investigational product group. Blinding minimizes the chance of treating the subjects differently based on the group to which they are assigned, minimizing bias from either the investigator or the subject. For example, if the investigator or coordinator knows that a subject is assigned to the investigational product group she or he may treat that subject differently than a subject assigned to placebo. The study group includes the individuals who receive the investigational product, and the control group receives either standard of care or a placebo. Subjects are randomized into the study groups by chance. Randomization is similar to a coin toss that determines the group to which the subject will be assigned. The randomization process is included in the research design and statistical analysis plan that has been created by qualified biostatisticians. Assignments are generally determined by a computer generated randomization program created to the specifications of the biostatisticians. During Visit 1, a subject is enrolled in the study and randomized into either the study or control group.

As discussed previously in this chapter, the protocol guides the research site as to what tests and procedures need to occur during each visit. All scheduled procedures are completed and the results are entered either on a paper case report form (CRF) or directly into an electronic database. If the results are entered on a paper CRF, these must then be entered into the electronic data capture (EDC) system in a timely manner. Results of the procedures at the first visit are referred to as baseline data. Data collected at future visits are compared to the baseline data to determine changes in the values.

At each visit, whether scheduled or unscheduled, the study coordinator must complete an AE and concomitant medication form. Adverse events are defined as undesired outcomes that occur during participation in the clinical trial. These events may or may not be related to the investigational product. (See sample AE form in Appendix B.) For example, a subject may report a toothache during a visit. This would need to be documented on the AE form, however it is likely not related to his or her participation in the study. In another case, a subject may report a stomachache, which would need to be documented and may or may not be related to his or her participation in the study. The investigator is responsible for determining the likelihood of whether or not a reported event is related to the investigational product and its severity (mild, moderate, severe). "An investigator must immediately report to the sponsor any serious adverse event (SAE), whether or not considered drug related, including those listed in the protocol or investigator brochure and must include an assessment of whether there is a reasonable possibility that the drug caused the event" (FDA IND Investigator Reports, 2015).

When documenting an AE, the following information must be captured:

- Event
- Start date
- Resolution date
- Treatment
- Severity
- Investigator's determination whether the event is possibly related to the investigational product

Other events that need to documented and reported include unanticipated AEs and SAEs. These are covered later in this chapter.

During the screening visit, a concomitant medication form, generally provided by the sponsor, needs to be completed. A member of the research team, typically, the study coordinator, asks subjects for a list of all drugs, including prescribed medications, over-the-counter drugs, supplements, and herbal products, they are currently using. Other medications, including over-the-counter medications and supplements, may impact the investigational product. These must be reviewed to determine if a medication/supplement will disqualify a potential subject. In addition, when an AE occurs, it could be due to a concomitant medication. These medications must be tracked throughout the study. At each visit, the coordinator reviews the concomitant medication form with the subject to determine if it is still correct or to identify and document any changes. (See Appendix B for a sample concomitant medication form.)

When subjects are randomized, they are also assigned to a specific number associated with the investigational or control product they receive throughout the study. Each time the subjects receive their assigned product, the numbers should be checked and double checked to ensure proper distribution of the assigned product (this process should be outlined in a formal standard operating procedure [SOP]). All investigational products must be stored in a secure area with limited access. Some investigational products may require storage in a refrigerator or freezer, and in these cases, a member of the research team needs to track the refrigerator or freezer temperature using designated and calibrated thermometers and then document the temperatures in temperature logs. How often this needs to be checked and recorded will be included in drug storage instructions that should be included with the drug. If the study includes this type of log, the information should be documented in a pharmacy manual.

Depending on the protocol and the manner in which the investigational product is dispensed, a member of the research team will dispense the investigational product assigned to the subject and log this on the investigational product accountability form. (See a sample investigational product accountability form in Appendix B.) To ensure accurate accountability of all investigational and control products, subjects are asked to return drugs not used and all packaging materials when they receive their next disbursement of drug. Not only does this allow for accurate drug accountability, it also provides data on subject compliance.

Some drugs are administered to the subject at the visit. These may include injections or intravenous infusions. If these drugs need to be mixed or prepared, this generally requires an unblinded nurse/pharmacist (depending on protocol requirements) to prepare the drug for dispensing.

Study visits must occur during protocol defined visit timeframes/windows. For example, if a visit is scheduled 30 days after the previous visit, a visit window might be 2 days prior or after the 30th day. Visit windows are determined based on the study and can be found in the study protocol.

Occasionally, a subject may miss a visit or a subject might report an AE that requires follow-up. These types of situations result is unscheduled visits. The protocol defines what must occur at these types of visits. However, if the visit is to follow-up on an AE, the investigator may order tests or procedures to ensure that it is safe for the subject to continue in the study.

Unscheduled Visits

During the conduct of a research study, there are multiple situations that may require the subject to come to the research site for an unscheduled visit. Some of the reasons for an unscheduled visit might be the following:

- The subject missed the visit window.
- The subject experienced an AE and needed to be seen for follow-up.
- The sponsor may amend the protocol and add extra visits.
- Due to lab results, the investigator has determined that for the welfare of the subject, she or he should be seen more often.

The data from these visits must be recorded just as any study visit. If a visit window has been missed, the visit might be rescheduled before the next scheduled visit. For this type of visit, the missed visit procedures would be performed (depending on the protocol limitations for making up missed visits). This may or may not be allowed per protocol.

Data collected from other unscheduled visits may or may not be similar to the regularly scheduled visits. Depending on the circumstance, such as an AE, the investigator will determine what types of procedures and services need to be conducted. If the protocol is amended and additional visits are added, the sponsor generally revises the visit schedule and provides the additional CRFs for these visits.

Of note here, unscheduled visits should be covered in both the study budget and the clinical trial agreement. These are covered in Chapters 7 (Budgets) and 8 (Contracts).

Final Study Visit

Subjects' final visit is similar to their first visit when baseline data were collected. The baseline tests that occurred during the first visit are usually repeated at the final study visit. The intermittent and final results of tests and procedures are compared with the baseline results to determine whether the investigational product has achieved the expected results. The data are also needed by the sponsor to prove that the investigational product is safe and effective in its application to the FDA for approval to market the product.

If a subject decides to withdraw from the study, it is important that he or she be brought in for a final visit in order to safely withdraw him or her from the study and to collect any unused investigational product. Occasionally, a subject may not contact the research site to drop out and may just stop coming in for visits. This subject would be considered "lost to follow-up." The research team should make multiple reasonable attempts to get the subject to come to the research site for a final follow-up visit. Additionally, the site SOPs should have a policies and procedures in place for handling subjects who are lost to follow-up, detailing how the research team will try to contact subjects and expected number of calls or attempts to contact before removing subjects and marking them as "lost to follow-up."

ADDITIONAL STUDY EVENTS

During the study, subjects may experience events that need to be reported to the institutional review board (IRB) and sponsor.

Unanticipated Events

Prior to administering investigational products to humans, preclinical studies are conducted in laboratories and in animal studies. This information is summarized in the investigator's brochure (IB). The IB is continually updated to include first-in-human studies and additional human studies as they are completed, providing additional drug-related information. Information in the IB includes risks and benefits of the investigational product as well as AEs that have occurred during animal and human studies. The AEs listed in the IB are not considered unexpected and would not be unanticipated events. If a subject reports an AE that has not been previously reported in the IB, this is defined as an unexpected AE. As outlined in 21 C.F.R. §312.64(b), unexpected AEs must be reported by the investigator to the sponsor and IRB. Reporting unexpected events include the following:

- The investigator reporting the unexpected event to the sponsor.
- The sponsor is responsible for reporting the unexpected event to the FDA.
- The investigator is responsible for reporting the unexpected event to the IRB.

Sponsors are required to notify all participating study investigators and the FDA of "any adverse experience associated with the use of the drug that is both serious and unexpected" and "any finding from tests in laboratory animals that suggests a significant risk for human subjects" (FDA IND Safety Reporting, 2015). If the study has a data safety monitoring board (DSMB), the sponsor must also submit this information to it. A DSMB is an independent committee of qualified individuals who review the adverse, unanticipated, and serious events for the overall clinical trial to determine whether or not it is safe to continue conducting the clinical trial (U.S. FDA, 2006).

Serious Adverse Events

An SAE is defined as a death or as life threatening, requiring hospitalization, or causing disability or permanent damage or if exposure to the investigational product during pregnancy may have resulted in an adverse outcome in the child (FDA IND Safety Reporting, 2015). All SAEs should be reported promptly to the sponsor and the IRB. Follow-up reports are required and should follow the initial report in a timely manner. All SAEs require follow-up through resolution of the event. This may include obtaining records from hospitals, health care practitioners, and any professional who was directly involved in treating the subject for the SAE. The sponsor is required to report all SAEs to the FDA, complete a MedWatch form, and disburse the MedWatch reports to all investigators conducting the study. The investigator at each research site must review each MedWatch report and acknowledge the review by signature.

Unblinding

Occasionally, in order to properly treat and manage a subject's SAE, the treating physician must know whether the subject has been randomized to an investigational product or to a placebo. If this occurs, the investigator is allowed to follow the unblinding process provided by the sponsor at the beginning of the study. At the beginning of the study, the investigator is entrusted with sealed envelopes or secure access to a computer database that includes the information identifying the group to which the subject is assigned. If unblinding is necessary, the investigator should contact the sponsor prior to unblinding, if possible. If contact with the sponsor is not possible, the investigator must report the unblinding to the sponsor as soon as possible.

PROTOCOL DEVIATIONS

A protocol deviation is defined as "an unplanned excursion from the protocol that is not implemented or intended as a systematic change" (U.S. FDA, 2015, Section D3). Protocol deviation is also used when referring to other unplanned protocol noncompliance incidents (U.S. FDA, 2015, Section D3). A deviation does not increase risk or decrease the benefit and does not have a significant effect on the subject's safety or rights. A deviation may be the result of the research staff, the investigator, or the subject actions. Examples of deviations might be a coordinator forgetting to perform a required procedure at a visit; a subject not completing a visit during the set window due to a vacation or conflict; or an investigator stopping a procedure such as an intravenous administration of the investigational product due to a harmful subject reaction. All deviations must be documented and reported to the sponsor and the IRB. Deviations are generally included in the continuing review report that the investigator is required to submit to the IRB on a regular basis.

PROTOCOL VIOLATIONS

A protocol violation occurs as a result of a change to the study protocol or noncompliance with the approved protocol without prior approval of the sponsor and the IRB. Violations might increase risk, decrease benefit, and affect the rights, safety, or well-being of the subject. A violation may also affect the integrity of the data. Examples of violations might include failure to provide informed consent to the subject prior to conducting study procedures, administering the wrong drug or dose to a subject, or randomizing a subject who did not qualify for the study based on eligibility criteria. All violations must be reported to the sponsor and the IRB within the set number of business days in a format (fax, email, hard copy, etc.) designated by the sponsor and IRB.

PROTOCOL AMENDMENTS

Once a study protocol has been approved by an IRB, any subsequent change to it is considered a protocol amendment. Amendments must be submitted to an IRB and receive approval prior to implementation of the change. Most studies, especially long-term studies, have multiple amendments during the life of the study. For example, if there is a new AE reported that might affect the safety and welfare of the study subjects, both the protocol and the informed consent must be amended to reflect this change. When an informed consent is amended, each subject in the clinical trial must be reconsented.

CONTINUING REVIEW REPORTS

The investigator of a study is responsible for submitting regular study reports to the IRB. The IRB determines how often these are required (per IRB SOPs), generally every 6 or 12 months. The investigator must also submit a final report to the IRB after study closure. Information found in the continuing review reports might include number of subjects enrolled, gender, race, age, list of amendments, AEs, deviations, violations, and safety reports.

DOCUMENT MANAGEMENT

It is the responsibility of the investigator to provide complete and accurate data for a study. This includes making sure that all source data are accurately transcribed in the CRFs and entered into the EDC system. The sponsor's monitor also checks the study data for accuracy and completeness. Incorrect or incomplete data result in a data query from the monitor. Queries are resolved through the comparison of subject records and the correct data entered in to the EDC system.

According to International Conference on Harmonisation (ICH) E6 (R2), appropriate corrections, additions, or deletions that are made to CRFs must be dated, explained (if necessary), and initialed by the investigator or by a member of the investigator's trial staff who is authorized to initial CRF changes for the investigator (ICH, 2015). It is a generally accepted practice to strike through incorrect data with one single line (for example, ~~example~~) so the original information is still readable.

The individual responsible for changing the data initials and dates the change and provides the reason for the change if applicable.

ALCOA is a common acronym used in the discussion of quality data evidence in both electronic and written documentation. Coined by Stan Woollen in his role as Bioresearch Monitoring Program Coordinator at the FDA in the early 1990s, ALCOA describes the attributes of quality data collection in clinical research and is used as a quality indicator in FDA inspections (Woollen, 2010, p. 1). ALCOA is an acronym for the following:

- Attributable: It must be clear who created the record—data should be linked to its source.
- Legible: Reviewers must be able to read the data—data are readable.
- Contemporaneous: Data must be recorded as it occurs, which includes the current date and time of occurrences—data are recorded promptly.
- Original: Records must be original—source data as first recorded.
- Accurate: The information recorded is correct—free from error (Woollen, 2010, p. 3).

For data that are captured electronically, the electronic system must meet the requirements outlined in 21 C.F.R. 11, "Electronic Records, Electronic Signatures." The system must be secure with limited access via secure login and password. Personnel using the system must have a signature on file and sign each change via an electronic signature that has been assigned to that specific person. The system must track any change made to the data and identify the person who made the change, the date and time of the change, and the reason for the change. The FDA controls and requirements include the following:

- Limiting system access to authorized individuals
- Use of operational system checks
- Use of authority checks
- Use of device checks
- Audit trail of all user activities
- Determination that persons who develop, maintain, or use electronic systems have the education, training, and experience to perform their assigned tasks
- Establishment of and adherence to written policies that hold individuals accountable for actions initiated under their electronic signatures
- Appropriate controls over systems documentation (FDA Electronic Records; Electronic Signatures, 2015)

The FDA requires that all study records be stored for 2 years from the date the investigational product is approved for market (approved New Drug Application) or the study is terminated.

A VIEW FROM INDIA

The process of enrollment through final study visit is similar in India as it is in the United States, as outlined in this chapter. However, there are certain differences in the definitions of AEs, adverse drug reaction (ADR), and SAEs or drug reactions as found in the GCP Guidelines published by the Central Drugs Standard Control Organization (CDSCO).

- Adverse event:
  Any untoward medical occurrence (including a symptom/disease or an abnormal laboratory finding) during treatment with a pharmaceutical product in a patient or a human volunteer that does not necessarily have a relationship with the treatment being given.
- ADR:
  - In case of approved pharmaceutical products: A noxious and unintended response at doses normally used or tested in humans.
  - In case of new unregistered pharmaceutical products (or those products which are not yet approved for the medical condition where they are being tested): A noxious and unintended response at any dose(s).
  - In case of an ADR, there appears to be a reasonable possibility that the AE is related with the investigational product being studied.
- SAE or serious ADR:
  An AE or ADR that is associated with death, inpatient hospitalization (in case the study was being conducted on out-patients), prolongation of hospitalization (in case the study was being conducted on inpatients), persistent or significant disability or incapacity, a congenital anomaly or birth defect, or is otherwise life threatening (Government of India, Ministry of Health & Family Welfare, CDSCO, 2004).

AE and SAE Reporting

All unexpected AEs, ADRs, and SAEs should be reported to the sponsor by the investigator within 24 hours of occurrence and to the ethics committee (IRB/independent ethics committee [IEC]) that approved the study protocol within 7 days. In the event of death, the IRB/IEC should be informed within 24 hours. Any unexpected SAE as defined in the Indian GCP Guidelines occurring during a clinical trial should be reported within 14 calendar days by the Sponsor to the Licensing Authority (DCGI) and to the investigator(s) of other clinical trial sites participating in the study. All other serious unexpected reactions that are not fatal or life threatening must be filed not later than 14 calendar days. At the end of the trial, all AEs, whether related to the trial or not, must be listed, evaluated, and discussed in detail in the final study report (Indian Council of Medical Research New Delhi, 2006, p. 43).

Sponsors reporting SAEs follow the guidance given by the Drugs Controller General of India. Forms, such as CIOMS-I, MedWatch, or any other company-specific or self-designed forms that are required, must meet the reporting format described in Appendix XI of Schedule Y (CDSCO; Drugs Controller General (India), 2011, p. 6).

In a recent amendment to the Drugs and Cosmetics Act 1940 and the Drugs and Cosmetics Rules 1945, the detailed procedure for the analysis of SAEs (including deaths) is outlined. The procedures include information on how the cause of death/injury should be determined and how the amount of compensation to be paid by the sponsor to the subject or his nominee should be determined (Office of Drugs Controller General (India), 2014).

Protocol Amendments

There are typically two different types of amendments used in India, administrative modifications and clinical modifications.

Administrative or technical modifications (like a change in a study coordinator, nurse or monitor, or telephone or fax numbers), which do not interfere with the subject's health interests, must be in writing and filed as amendments to the protocol. All persons responsible for the study must sign these amendments. The ethics committee who approved the study should be informed by the principal investigator regarding this amendment.

Clinical modifications that interfere with a subject's health interests and involve changes in the design of a study or its scientific significance require a new approval by the ethics committee. All persons responsible for the clinical study must sign any such modification. The amended protocol and informed consent documents should be approved and the subjects should reconsent to continue participation in the study.

Document Management

As required by ICH GCP guidelines, an investigator must keep essential study documents for at least 2 years after the last approval of a marketing application in an ICH region and until there are no pending or proposed marketing applications in an ICH region or at least 2 years have transpired since the formal discontinuation of clinical development of the investigational product.

However, if required by applicable regulatory requirement(s) or if required by the sponsor, these documents should be retained for a longer period. An investigator must make provisions for subjects' medical records to be kept for the same period of time. Additionally, subjects' medical records and other original data should be archived in accordance with the archiving requirements of the investigational sites.

CHAPTER REVIEW

In this chapter, subject study visits were presented from the screening visit to the subject's final visit. The chapter defined what occurs during each visit. In addition, the chapter looked at other situations that occur or might occur during a study visit or during the duration of the study that must be addressed by the research site. These include unanticipated AEs, SAEs, deviations, violations, unblinding, amendments, monitor visits, continuing reviews, and document management.

APPLY YOUR KNOWLEDGE

1. On the Delegation of Authority form, the investigator has assigned authority to conduct informed consents to the study coordinator, research nurse, and research assistant. You are the assigned coordinator for a study and have a research assistant to help with the study. The research assistant

conducted the informed consent process with a potential study volunteer. The assistant reports to you that the consenting process has been completed and the individual has agreed to be screened for the study. You instruct the assistant to make a copy of the informed consent form and give it to the individual before she or he leaves. You take the individual into a exam room and perform all of the screening procedures. The research assistant provides a copy of the informed consent to the individual and you inform the individual that you will contact him or her when you have the results from the tests and procedures. You give the paper CRF and informed consent form to the data entry team to enter into the EDC system. At the end of the day, a staff member from data entry comes to your desk and informs you that while he was entering the data into the EDC, he noticed that the informed consent had not been signed by the potential subject. What actions should you take?

2. You have been assigned by the investigator of your site to evaluate and recommend a EDC system. What does the system need to include to be compliant with 21 C.F.R. 11?

REFERENCES

Drugs Controller General (India). (2011). *Guidance for industry on reporting serious adverse events occuring [sic] in clinical trials.* Retrieved from http://www.cdsco.nic.in/writereaddata/sae%20guidelines%2005-05-2011.pdf

FDA Electronic Records; Electronic Signatures, 21 C.F.R. § 11 (2015).

FDA IND Safety Reporting, 21 C.F.R. § 312.32 (2015).

FDA IND Investigator Reports, 21 C.F.R § 312.64 (2015).

Government of India, Ministry of Health & Family Welfare, Central Drugs Standard Control Organization. (2004). Good clinical practice for clinical research in India. Retrieved from http://www.cdsco.nic.in/html/GCP1.html

Indian Council of Medical Research New Delhi. (2006). *Ethical guidelines for biomedical research on human participants.* Retrieved from http://www.icmr.nic.in/ethical_guidelines.pdf

International Conference on Harmonisation (ICH). (2015). Integrated Addendum to ICH E6 (R1): Guideline for Good Clinical Practices E6 (R2).

Office of Drugs Controller General (India). (2014). *System of pre-screening for submission of reports of SAEs to CDSCO* (F. No. 12/01/13-DC (Pt-13A)). Retrieved from http://www.cdsco.nic.in/writereaddata/System%20of%20Pre-screening%20for%20submission%20of%20reports%20of%20SAEs%20to%20CDSCO.pdf

U.S. Food and Drug Administration. (2006). *Guidance for clinical trial sponsors: Establishment and operation of clinical trial data monitoring committees.* Retrieved from http://www.fda.gov/OHRMS/DOCKETS/98fr/01d-0489-gdl0003.pdf

U.S. Food and Drug Administration. (2015). Inspections, compliance, enforcement, and criminal investigations. Retrieved from http://www.fda.gov/iceci/enforcementactions/bioresearchmonitoring/ucm133569.htm

Woollen, S. W. (2010, summer). Data quality and the origin of ALCOA. *The Compass.* Retrieved from http://www.southernsqa.org/newsletters/Summer10.DataQuality.pdf

6 Collaborating for Compliance and Quality Data—Monitoring and Audits

The ultimate goal of drug and device development is to bring safe and effective medical products to the public. This requires dedicated expertise and oversight throughout the process to ensure that the "proof" of safety and effectiveness meets the standards of scientific integrity, while maintaining ethical principles. During the trial clinical process, regulations require sponsors (or their representatives) to monitor research sites for quality data and regulatory compliance. At the same time, the research team may conduct internal quality management reviews to ensure continuous quality data standards, and the (approving) institutional review board (IRB), the Food and Drug Administration (FDA), and other internal and external organizations may audit research sites for scientific, ethical, and/or regulatory reasons.

In this chapter, we explore the functions of monitoring and auditing and the roles of the sponsor, the investigator, the IRB, the FDA, and other organizations in impacting quality and compliance at the research site.

MONITORING VERSUS AUDITING

What is the difference between monitoring and auditing, and what significance does this have in the oversight of a clinical trial? The International Conference on Harmonisation (ICH) defines the two functions in the following manner:

- Monitoring is "the act of overseeing the progress of a clinical trial, and ensuring that it is conducted, recorded, and reported in accordance with the protocol, standard operating procedures (SOPs), GCP, and the applicable regulatory requirement(s)" (ICH, 1996, Section 1.38).
- Auditing is "a systematic and independent examination of trial-related activities and documents to determine whether the evaluated trial-related activities were conducted, and the data were recorded, analyzed, and accurately reported according to the protocol, sponsor's standard operating procedures (SOPs), GCP and the applicable regulatory requirement(s)" (ICH, 1996, Section 1.6).

Although they seem similar in intent and purpose, monitoring and auditing are conducted from two different perspectives and serve distinctive purposes in the conduct of clinical trials. Monitoring is an ongoing quality control (QC) process of routinely and regularly assessing clinical trial conduct from trial initiation through trial termination. Conversely, an audit is conducted at a single point in time to assess the current state of or a single aspect of a clinical trial. Auditors offer an independent opinion of clinical trial activities and report their findings through official reporting mechanisms, such as inspection findings, determination letters, and other types of reports. Monitoring is analogous to a video that takes place over a particular timeframe, whereas auditing is similar to a snapshot taken at one specific time.

In clinical research, it is not unusual for a sponsor to monitor and audit clinical studies as part of its responsibility of clinical trial oversight. Similarly, a research site may conduct internal monitoring and may also invite an independent entity to audit various aspects of a clinical trial. For example, a hospital-based, cardiac research program may collaborate with another research program within the same hospital to conduct an internal data quality inspection audit to gain a unique perspective on a clinical trial.

SPONSOR MONITORING

The sponsor is responsible for ensuring that clinical investigations are properly monitored and that they are conducted per the general investigational plan and per the protocol found in the Investigational New Drug or Investigational Device Exemption application. Clinical trial monitoring is conducted to verify that "the rights and well-being of human subjects are protected; the reported trial data are accurate, complete, and verifiable from source documents; and the conduct of the trial is in compliance with the currently approved protocol/amendment(s), with GCP, and with the applicable regulatory requirement(s)" (ICH, 1996, Section 5.18.1).

As part of the monitoring function, the sponsor is responsible for selecting monitors who are qualified by training and experience to observe the progress of an investigation. The monitor may be an employee of the sponsor, a contractor, or another individual meeting the training and experience qualifications of a particular clinical trial. She or he is familiar with the investigational product, the protocol, the informed consent form, case report forms, standard operating procedures, and good clinical practices (GCPs)/regulations. In most cases, the monitor also has knowledge and experience within particular disease categories. For example, a monitor with a strong oncology background would understand the nuances of data collection across various chemotherapy and radiation treatment modalities.

Prior to the start of a clinical trial, the sponsor and clinical trial monitor, in keeping with sponsor guidelines and standard operating procedures (SOPs), develop a monitoring plan that defines how a research site will be monitored, the frequency of monitoring visits, and the activities that will be performed at visits. Depending on the phase of the clinical trial, the experience of the research team, the complexity of the clinical trial, and the disease being evaluated, the monitoring plan may include review of 100% of the data for all subjects enrolled. Monitoring plans for later stage, less complex studies with investigational products that have strong safety profiles may vary greatly in the timing and percentage of data monitored.

In 2013, the FDA published a guidance document recommending a quality risk management approach to the monitoring of clinical trials, otherwise known as risk-based monitoring (RBM). The goal of RBM is to increase monitoring efficiency by targeting monitoring activities on the most critical data in preventing likely sources of error in conducting, collecting, and reporting clinical trial data, while maintaining subject safety and data quality. The FDA encouraged sponsors to use a monitoring approach based on the specific risks for each clinical trial, concentrating resources where they would deliver the greatest benefit to the clinical trial rather than using a broad monitoring plan where resources would be overextended. In other words, the FDA suggested that monitoring resources would be best served by concentrating the resources where they would add quality/value to clinical trial oversight.

To employ RBM, many sponsors have adopted the assumptions of the TransCelerate™ BioPharma RBM Methodology, which include

- Central and offsite monitoring;
- Targeted monitoring activities;
- Methodology tailored to available technology;
- Sponsor expectations for data entry and query resolution;
- Functional oversight amended to meet changes in risk;
- RBM expectations documented as SOPs;
- RBM applied to all phases, types, and processes of clinical trials;
- Communication tailored to effectiveness; and
- Risk assessments implemented before finalization of protocols and case report forms (TransCelerate Biopharma Inc., 2013, p. 6).

In summary, the RBM methodology requires a real-time, adaptive methodology of utilizing risk factors and communication to identify, assess, plan, track and control risk.

Sponsor Monitoring Visits

At the writing of this book, RBM is still being implemented by many sponsors, and as a result, monitoring visit frequency, timing, and methodology are in transition. However, typical monitoring visits are conducted during preclinical trial start-up, during the conduct of the clinical trial (interim visits), and during clinical trial close out (close-out visit). Each type of visit serves a specific purpose, requiring the monitor to interface with the investigator and his or her research team. As mentioned previously, depending on the sponsor, the type of clinical trial, and the research site, monitoring visits will vary. Not every clinical trial will have all three types of monitoring visits, and many monitoring visits may be conducted remotely. The following paragraphs describe typical monitoring visits.

Preclinical trial visits include research site qualification visits and site initiation visits. It is during the research site qualification visit that the monitor meets with the investigator and his or her research team, tours and inspects the facility, assesses and documents the source documentation/collection method, and collects and reviews any outstanding documents. In essence, the monitor makes sure that the investigator understands the clinical trial, has adequate resources, and that processes are in place to collect and document clinical trial data. During the research site initiation evaluation, the monitor focuses his or her efforts on training, educating, and explaining various aspects of the clinical trial and reminding investigators of their responsibilities in delegation, informed consent, source data, and other specific clinical trial requirements.

Interim visits are conducted on a routine basis per the monitoring plan to

- Review serious adverse event (SAEs) and their reporting status.
- Ensure all subjects that have been screened and/or enrolled have signed an informed consent form, and that they have met the inclusion/exclusion criteria.
- Ensure any protocol violations have been addressed and reported.
- Ensure that the research team and site facilities remain within GCP guidelines and expectations.
- Review data to ensure that they are accurate and complete and to discuss queries.

On the whole, interim visits are conducted to monitor the progress of a clinical trial and to maintain a communication link with the research team. At the end of a visit, the monitor discusses his or her findings with the investigator and research team, underscoring any areas of concern that might require a corrective action plan. To document any findings further, the monitor submits a site visit report to the sponsor listing all findings at the research site. The findings are summarized in a follow-up letter to the investigator at the research site, who should address any findings and any proposed corrective actions (and a timeline) in a written communication to the sponsor to resolve any deficiencies or concerns.

At the end of a clinical trial, there are clinical-trial-related details that need to be resolved and rectified, such as determining timelines for finishing outstanding queries. During the close-out visit, the monitor resolves outstanding items from the interim visits and then focuses on the reporting and regulatory requirements that are required for the clinical trial to close. In particular, she or he discusses the obligations and disposition of data, investigational products, clinical trial supplies, and other trial-related materials.

The Research Site Perspective

Some of the attributes of RBM such as centralized monitoring, targeted onsite monitoring, and electronic transfer/access to research records have had a confusing impact on research sites. The resource needs to support the administrative requirements of remote document access have increased; the methods, timing, and content of communications have changed; and the planning, resourcing, and training of research staff have transitioned to accommodate the changes in technique and implementation. Typical preparation for monitoring visits, such as finding a quiet place for the monitor to review data, scheduling appointments for the monitor to meet with the pharmacist, and communication between the monitor and investigator, has evolved to accommodate electronic and telephonic monitoring visits, offsite access to data, and data transfer methods. Consequently, it is important for the sponsor and the research team to have clear expectations of each other that are defined in the clinical trial agreement as well as in the monitoring plan and throughout ongoing communications. If research site personnel are aware of expectations, timelines, and communication methods, they will be much more likely to be compliant and continue to provide the sponsor with access to quality data that meet the overall goals of the clinical trial.

INTERNAL QUALITY MANAGEMENT

Before and during the course of a clinical trial, there are numerous opportunities for research teams to institute quality management mechanisms that increase the accuracy and timeliness of data, which can have a direct impact on a research site's financial bottom line. Quality management processes, such as quality assurance (QA) and control, are aimed at identifying and preventing errors, deficiencies, or other compliance and safety concerns, as well as correcting any errors that are identified. A quality research site is very appealing to a sponsor that expects a clinical trial to be conducted accurately and efficiently while maintaining subject safety and regulatory compliance.

QA is a set of activities that are put in place to establish requirements and procedures, to ensure that the requirements are being met and procedures are being followed, and to verify that standards are maintained. Two QA processes discussed throughout this book include implementation of research site staff training, both at the GCP level and at specific clinical trial procedure levels, and implementation and integration of SOPs. QA audits are conducted to determine if documentation adheres to written procedures, policies, and regulations. An example of a QA activity would include the review of a training log to ensure that sufficient and relevant training has been completed and documented. Internal QA audits should be conducted on a regular basis and by someone other than the person responsible for implementing the procedures/processes being reviewed. In many cases, research site managers hire a consultant to conduct an audit to minimize bias and to offer an independent report.

QC is a real-time set of operational activities intended to verify that data are generated, collected, analyzed, and reported according to SOPs, GCPs, and the research protocol. Examples of QC activities include completion of checklists, ongoing review of Case Report Forms to source documentation, temperature log maintenance for clinical trial freezers and refrigerators, equipment calibration checks, and other activities that support the quality of the clinical trial. QC procedures should be defined in the SOPs and should be conducted on a regular schedule. QC is typically conducted by research site coordinators and/or a designated quality specialist at the site.

Ideally, all clinical trials should have a quality plan in place prior to initiation of the trial at the research site. In many grant-funded research studies, a clinical quality management plan (CQMP) to facilitate the quality execution of specific studies is required by the funding agency. The CQMP is a written document that details the ongoing processes, responsibility, scope, indicators measured, and frequency of the activities within the protocol. Some of the key quality indicators measured in a CQMP include eligibility criteria, informed consent process, missed visits, SAEs, and concomitant medication documentation. Tools and templates for developing and implementing CQMP are available to the public through the National Institutes of Health (NIH) website or other federal and state research websites.

CORRECTIVE ACTION/PREVENTIVE ACTION

Corrective action/preventive action (CAPA) plans are part of quality management and are a means of identifying a problem and seeking preventive action to keep the problem from recurring. The International Organization for Standardization (2015, Section 3.12) defines CAPA as "corrective action: action to eliminate the cause of a nonconformity" and "preventive action: action to eliminate the cause of a potential nonconformity." In other words, corrective actions are taken to resolve a problem and preventive actions are taken to keep the problem from recurring. Although CAPA originated in the electronics industry and is a part of good manufacturing processes, it is highly adaptable to resolving problems in other industries. Corrective action/preventive action has been used by investigators for years, but it is now a general expectation that CAPAs are more formally documented, implemented, and evaluated over time for effectiveness. The FDA has demonstrated its strong support of quality management implementation and CAPA processes throughout drug and device development processes and has published guidelines and training modules on its website.

ROOT CAUSE ANALYSIS

To adequately write a CAPA plan, the cause of a problem needs to be identified. A root cause analysis (RCA) is a problem-solving method used to identify the root of a problem, rather than just the obvious symptoms. The "5 Whys" method can help the user find the root cause of a problem. This is done by stating the problem, then asking "why" at least five times, or until the underlying source of the problem is revealed.

The 5 Whys method is used in the following example to find the root cause of a problem identified during a monitoring visit:

- Protocol: Protocol 539 requires subjects to have a complete blood count (CBC) conducted at a designated clinical trial laboratory 1 hour prior to the first administration of the investigational isotope.
- Problem: Subjects 18, 16, 12, and 21 had CBC laboratory tests *less* than 1 hour prior to administration of the investigational isotope.
  - Why did the subjects have blood draws less than 1 hour prior to administration? *Subjects did not go to the study-specific laboratory for their blood draw. Instead, they went to a "nonstudy" laboratory.*
  - Why did the subjects go to a laboratory that was not participating as a provider in the clinical trial? *The clinical trial instructions for obtaining blood draws included the name of a laboratory that was not part of the clinical trial.*

- Why did the instructions for obtaining blood draws include the name of a laboratory that was not part of the clinical trial? *The instructions were not updated prior to the start of the clinical trial.*
- Why weren't the instructions updated prior to the start of the clinical trial? *The new clinical research assistant did not know that he was responsible for updating clinical trial instructions.*
- Why didn't the clinical research assistant know that he was responsible for updating clinical trial instructions? *The research site SOPs did not clearly define who is responsible for updating clinical trial instructions.*

In the above case, the underlying source, or root cause, of the problem was the research site's SOPs. The SOPs did not clearly allocate responsibility for updating clinical trial instructions. How will the problem be fixed? And how can the problem be prevented in the future? This should be answered in the CAPA plan.

CAPA Plan

A typical CAPA plan is in memo format and includes the following elements: date, issue/concern, root cause, description of the corrective actions, who is responsible for the corrective actions and timeline, documentation of the procedures used to resolve the problem, effective date of resolution, description of the preventive actions planned, plans for follow-up, and any additional comments. An example of a CAPA plan for the previous root cause example follows:

Date: June 18, 2016
To: Good Isotope Sponsor
From: Dr. Great Investigator, Principal Investigator for GIS Isotope Clinical Trial
Issue: Protocol 539 requires subjects to have a CBC conducted at a designated clinical trial laboratory one hour prior to the first administration of investigational isotope. Subjects 18, 16, 12, and 21 had CBC laboratory tests *less* than 1 hour prior to administration of the investigational isotope.
Root Cause: Subjects 18, 16, 12, and 21 did not have their laboratory test completed 1 hour prior to administration of the clinical trial isotope because they went to the laboratory listed in the clinical trial instructions instead of the laboratory designated for this clinical trial. The clinical trial instructions had not been updated from a previous clinical trial because the assistant did not realize it was his responsibility to do so. The root cause of this issue is the lack of clarity of our research site SOPs in regard to clinical trial instruction updates and responsibility.
Corrective Action: Upon discovery of the root cause of this issue (2 weeks ago), the research site manager immediately revised the Clinical Trial Instruction SOPs to clearly define the responsible party for and timeline of updating the Clinical Trial Instructions prior to trial implementation. Additionally, the clinical research assistant was advised of this and was retrained on his responsibilities by the lead clinical trial coordinator.

Implementation: Upon discovery of the root cause of this issue, the research site manager supervised the updating of clinical trial instructions by the research assistant, and we (research site manager and I) implemented a standard footer on clinical trial instructions with the date of the update and initials of person completing the instruction update (this is also now defined in the SOPs).

Effective Date of Resolution: June 4, 2016 (date of discovery of root cause)

Preventive Action: Upon review of the "Clinical Trial Instruction" SOP, the research site manager noticed that many of our other SOPs also did not have clear lines of responsibility for completing and reviewing tasks. The clinical trial manager and lead coordinator updated our first SOP, "SOP for SOPs," to include lines of responsibility and reviewing tasks, and as a team, the research site manager and coordinators are updating each of our subsequent SOPs to conform to the "SOP for SOPs" with clear lines of responsibility, and second review on all tasks. The first draft of the SOPs will be completed by August 1, 2016, and review of the SOPs will be conducted by the lead coordinator and me by August 31, 2016.

Follow-up: Because SOPs are a key part of our quality research site, we will be evaluating their relevance and effectiveness on a quarterly basis, rather than annually. Additionally, we are "featuring" one SOP each week during our regular staff meetings to gain continuous quality feedback from the research team.

Comments: Thank you for the opportunity to respond to your request for a CAPA plan.

Great Investigator, M.D.	June 18, 2016
PI Signature	Date of Signature

Great Investigator, MD
PI Printed Name

In the previous examples, the root of the problem was identified as the SOP; however, to be realistic, we need to recognize that there may be multiple roots that impact a problem. Perhaps one of the research subjects overslept and missed his appointment for the blood draw? Or the research assistant was not paying attention when he was told to update the clinical trial instructions? We need to be aware that the 5 Whys method is a straightforward method of finding a root cause, but it is linear in nature and has limitations. A CAPA plan may need to include multiple courses of action to correct and prevent future incidents.

AUDITS AND INSPECTIONS

Because clinical trials are overseen by a variety of regulatory bodies, sponsors, and institutional officials, they are subject to audits by these same authorities. Some of the regulatory bodies, such as the FDA, refer to audits as "inspections" and have specific terminology associated with the conduct and follow-up of the audit. Similarly, the Office for Human Research Protections (OHRP) refers to audits as "compliance oversight evaluations" and has specific procedures in place. Other auditing bodies may use different procedures and terminology, but the purposes and goals of their audits are similar.

Although we are focusing on FDA inspections in this section, other agencies and entities are likely to audit research sites and/or studies. A few of these entities and organizations and their purpose for auditing research sites follow:

- Clinical trial sponsor—A sponsor audits a research site to ensure that subjects' rights and safety have been maintained, to ensure the integrity and reliability of the data, and to ensure that the clinical trial is compliant with the investigational plan, protocol, SOPs, and regulatory and GCP standards. Audits are done routinely as a best practice to evaluate sponsor policies and monitoring processes, and to prepare for the New Drug Application to the FDA. Additionally, audits are conducted in anticipation of an FDA inspection. A sponsor provides the research site with a written report within several weeks after the audit. Many sponsors classify observations as (1) Critical, meaning that there are significant issues impacting the reliability and integrity of the data or that present significant risk to subjects (possible systems failure). (2) Major, meaning that an error suggests significant compliance deficiencies with GCP or the research site's SOPs. And (3) Minor, meaning that an error suggests some deficiencies in compliance. Sponsors require a response to the written report and often request the research site to provide a CAPA plan.
- IRBs—An IRB may choose to audit a research site to verify that the clinical trial that it has approved is being conducted per its approval and to ensure proper documentation, data collection, and adherence to regulations, GCP, and IRB policy. Reasons a clinical trial may be audited include allegations of noncompliance, sponsor concerns, potential high risk to subjects, recruitment of vulnerable populations, research outside of investigator expertise, safety, or other reasons at the IRB's discretion. Should the IRB find specific compliance issues, institutional and IRB policies dictate consequences, including reporting, and corrective actions.
- Clinical trial institutions—Many research sites are found within a hospital system, academic center, or other institution that provides health care or other services. The institution's compliance office typically oversees auditing of the clinical trial processes, procedures, and operations within the institution to ensure that the research site is compliant with accreditation standards, billing and Medicare regulations, and other regulatory standards specific to the overall institution. Institutional compliance auditors also verify that the research site is compliant with GCP and regulatory compliance.
- Office for Human Subject Protections (OHRP)—OHRP oversees compliance of institutions and human subjects' research that falls under the jurisdiction of 45 C.F.R. 46. The agency conducts both for-cause and not-for-cause compliance oversight evaluations. For-cause evaluations are conducted in response to written allegations or indications of noncompliance. Alternatively, not-for-cause evaluations are conducted based on the following indications: volume of research conducted or supported by the U.S. Department of Health and Human Services, history of a relatively low level

of reporting to OHRP, need to evaluate corrective actions following a for-cause evaluation, geographic location, status of accreditation by a professional human subject protection accrediting organization, and/or status of recent human subject protection evaluations or audits by other regulatory agencies (OHRP, 2009). Following an evaluation, the OHRP sends a letter of determination to an institution detailing concerns and recommendations. Possible outcomes include the following:

- No identification of noncompliance;
- OHRP recommends improvements to institutions human subject protection policies;
- OHRP requires corrective action plans;
- OHRP restricts its approval of an institution's Assurance;
- OHRP suspends an institution's Assurance;
- OHRP recommends to HHS that an institution or investigator be temporarily suspended or permanently removed from participation in specific projects;
- OHRP recommends to HHS that institutions or investigator be debarred (government sanction); and
- OHRP refers the matter to another Federal department or agency for further review and action (OHRP, 2009).

- NIH—The NIH is a grouping of 27 different federally funded institutes and centers dedicated to science and health research. Research conducted through sponsorship from the NIH is subject to audits through the center or institute that sponsored the research. As an example, research sites that are participating in a National Cancer Institute clinical trial are subject to audit at least once every 3 years through the Clinical Trials Monitoring Branch. Institutions are required to submit CAPA plans for noncompliance and are subject to clinical trial suspension, probation, or withdrawal from participation in studies for noncompliance.
- Office for Civil Rights (OCR)—The OCR is part of the Department of Health and Human Services and is responsible for oversight of Health Information Portability and Accountability Act (HIPAA) compliance. The OCR initiates reviews in response to complaints, and it works with the Department of Justice (DOJ) to investigate possible criminal HIPAA violations, such as knowingly disclosing personal health information (PHI) (1 year prison sentence), using false pretenses to obtain PHI (5 year prison sentence), and disclosing PHI for marketing purposes (10 year prison sentence).
- Office of the Inspector General (OIG)—The OIG is tasked with protection of the integrity of HHS programs as well as the health and welfare of the beneficiaries of the programs, especially the most vulnerable populations. The OIG oversees the Centers for Medicare & Medicaid Services, Administration for Children and Families, Centers for Disease Control and Prevention, FDA, and NIH. Enforcement efforts are typically coordinated with the DOJ and include criminal and civil enforcement, as well as civil monetary penalties. The OIG concentrates the majority of its resources on oversight of Medicare and Medicaid.

FDA INSPECTIONS

The FDA created the Bioresearch Monitoring Program (BIMO) in 1977 to help ensure the protection of the rights, safety, and welfare of human research subjects; to verify the accuracy and reliability of clinical trial data in support of marketing applications; and to assess compliance with statutory requirements and FDA's regulations governing the conduct of clinical trials (U.S. FDA, 2010). The BIMO has responsibility for inspecting clinical investigators, sponsors, monitors, contract research organizations, IRBs, non-clinical (animal) laboratories, and bioequivalence analytical laboratories.

The FDA conducts inspections, as either announced or unannounced, under the following conditions:

- to verify the accuracy and reliability of the data that has been submitted to the agency;
- as a result of a complaint to the agency about the conduct of the clinical trial at a particular research site;
- in response to sponsor concerns;
- upon termination of the research site;
- during ongoing clinical trials to provide real-time assessment of the investigator's conduct of the trial and protection of human subjects;
- at the request of an FDA review division; and
- related to certain classes of investigational products that FDA has identified as products of special interest in its current work plan (i.e., targeted inspections based on current public health concerns) (FDA, 2010, p. 3).

Upon arrival and before an audit begins, an FDA inspector should present his or her identification/credentials and a Form FDA 482, "Notice of Inspection" for signature by the principal investigator (PI) or his or her representative. Once the form is signed, the inspector may ask for a tour of the facility and may wish to talk to various individuals involved with the clinical trial. Additionally, she or he will request various documents for review. Typically, the inspector will want to

1. Determine if the PI was in control of the clinical trial and if delegation was appropriate;
2. Determine if facilities were adequate for the clinical trial;
3. Determine if protocol was followed;
4. Know how and where data were recorded/stored;
5. Audit clinical trial files for compliance to regulations and GCP;
6. Determine sponsor and monitor interaction (corrective actions?);
7. Audit informed consent forms and data for evidence of subject safety and validity of data;
8. Determine investigational agent accountability; and
9. Whether financial interests were disclosed (FDA, 2010, p. 4).

Once she or he has completed an inspection, the FDA inspector conducts an exit interview where she or he discusses the findings of the inspection. If she or he has found deficiencies, the inspector issues a written Form FDA 483, "Inspectional Observations," to the investigator. Form FDA 483 describes observations that represent deviations from applicable statutes and regulations. The investigator has the opportunity to respond verbally to the observations or elect to respond in writing.

Common deficiencies include the following:

- Failure to follow the protocol and signed investigator statement
- Protocol deviations and reporting
- Failure to ensure informed consent was obtained in compliance with 21 C.F.R. 50
- Inadequate recordkeeping
- Inadequate accountability for the investigational product
- Inadequate subject protection, including IRB and informed consent issues (FDA, 2010, p. 7).

Outcomes of the inspection are classified into the following areas:

- NAI: No action indicated: No findings requiring action.
- VAI: Voluntary action indicated: Some findings noted, and the investigator should attempt to address the situation and respond in writing to the FDA district office.
- OAI: Official action indicated: Serious discrepancies noted. The investigator must rectify the situation and respond in writing to the FDA district office.

Per the FDA Information Sheet (2010), after supervisory review and classification of inspection findings, one of the following letters is usually sent to the investigator:

1. A letter that generally states that FDA observed basic compliance with pertinent regulations.
2. An Informational or Untitled Letter that identifies deviations from statutes and regulations that do not meet the threshold of regulatory significance for a Warning Letter. Such letters may request a written response from the clinical investigator.
3. A Warning Letter that identifies serious deviations from applicable statutes and regulations. A Warning Letter is issued for violations of regulatory significance and may lead to enforcement action if not promptly and adequately corrected. Warning Letters are issued to achieve voluntary compliance and include a request for correction and a written response to the agency.
4. Notice of Initiation of Disqualification Proceedings and Opportunity to Explain (NIDPOE). An NIDPOE is issued when the PI has repeatedly or deliberately failed to comply with the requirements for conducting clinical trials and/or has repeatedly or deliberately submitted false information to FDA or to the sponsor. The FDA may initiate a process to disqualify the clinical investigator from receiving investigational new drugs and/or biologics (FDA, 2010, pp. 7–8).

Of all the audits and inspections discussed in the previous pages, the FDA inspection is particularly important for industry-sponsored studies. Keeping in mind that the investigational product is under review for eventual marketing, it is clear that FDA inspections are warranted and needed to make sure that products meet the threshold and balance of safety and efficacy.

A VIEW FROM INDIA

As in the United States, clinical trials in India require dedicated expertise and oversight to meet scientific, ethical, and regulatory standards.

SPONSORS, AUDITS, AND INSPECTIONS

In India, the sponsor's responsibilities and purposes for monitoring the conduct of a clinical trial are similar to those in the United States. Additionally, processes for site selection, site visits, site QA and QC, RCA, and CAPA processes are comparable with those in the United States. However, unlike the United States, the terms *audit* and *inspection* are synonymous in regard to process.

As per the GCP guidelines published by Central Drugs Standard Control Organization (CDSCO), the audit of a clinical trial is defined as: "A systematic verification of the study, carried out by persons not directly involved, such as:

1. Study related activities to determine consistency with the Protocol,
2. Study data to ensure that there are no contradictions on Source Documents. The audit should also compare data on the Source Documents with the interim or final report. It should also aim to find out if practices were employed in the development of data that would impair their validity, and
3. Compliance with the adopted Standard Operating Procedures (SOPs)" (Government of India, Ministry of Health & Family Welfare, CDSCO, 2004, Section 1).

The CDSCO is the regulatory body, which in 2010 started the clinical trial inspection program and released a guidance document for clinical trial inspections, in India. The objective of the program is to verify GCP compliance to protect the rights, safety, and well-being of clinical trial subjects; to verify the authenticity and integrity of clinical trial data; and to verify compliance with various regulatory provisions as per the Drug & Cosmetics Act 1940 & Drug and Cosmetic Rules 1945 of India. Because the CDSCO collaborated with the U.S. FDA, and the inspectors were trained at workshops conducted by the U.S. FDA authorities, the processes of planning, conducting, and reporting inspections are similar in India to those in the United States. In addition to conducting inspections at clinical trial sites, they can also be carried out at the clinical research organization (CRO's)/sponsor's premises (Government of India, Ministry of Health & Family Welfare [MOHFW], 2010).

The Drug Controller General India (DCGI), who heads the CDSCO and has jurisdiction to inspect all clinical trials and his or her team of drug inspectors, can walk into clinical trial sites, at short notice, to conduct surprise and random audits of both the internal processes and the data of clinical trials. As per the Gazette of India Notification, Part II, Section 3, Subsection i, published on July 17, 2012, Rule 122 DAC (h), the premises of the sponsor/CRO and clinical trial sites shall be open to inspection by the officers of the CDSCO, who may be accompanied by an officer of the concerned State Drug Control Authority to verify compliance to the requirements of Schedule Y, GCPs guidelines and other applicable regulation (Government of India, MOHFW, 2012). If a research site/investigator is found to be in violation or noncompliant with guidelines, the DCGI could suspend operations at the site.

An inspection may be carried out for cause or not for cause. The clinical trials selected for inspection may be based on, but not limited to, the nature of the clinical study, complaints, data irregularities or regulatory decision based on clinical trial data, vulnerable study population, number of clinical trials, and subject enrolment at a particular trial site (MOHFW, 2010).

CHAPTER REVIEW

In Chapter 6, we concentrated our discussions on monitoring and auditing functions as quality processes during the course of clinical trials. Monitoring is a real-time process of overseeing activities and auditing is a snapshot of processes at a specific time. Monitoring is typically a QC process and auditing is a QA process. Corrective action/preventive action plans can assist research sites in developing plans to correct deficiencies discovered during monitoring and auditing visits.

APPLY YOUR KNOWLEDGE

1. The follow-up letter from a recent monitoring visit notes that Subject 16 did not receive a computed tomography (CT) scan after Visit 2, required by the protocol. Using the 5 Whys method, suggest the root cause of why Subject 16 did not receive a CT scan after Visit 2.
2. Upon determining a root cause for subject #16 not receiving a CT scan, write a CAPA plan directed to the sponsor.

REFERENCES

Government of India, Ministry of Health & Family Welfare, Central Drugs Standard Control Organization. (2004). Good clinical practice for clinical research in India. Retrieved from http://www.cdsco.nic.in/html/GCP1.html

Government of India, Ministry of Health & Family Welfare. (2010). *Guidance on clinical trial inspection*. Retrieved from http://www.cdsco.nic.in/writereaddata/CT%20Inspection%20-11-2-2011.pdf

Government of India, Ministry of Health & Family Welfare. (2012). *The gazette of India: Extraordinary, notification*. Retrieved from http://cdsco.nic.in/writereaddata/GSR%20572%20%20(E).pdf

International Conference on Harmonisation. (1996). *Guideline for good clinical practice E6(R1)*. Retrieved from http://www.fda.gov/downloads/Drugs/GuidanceCompliance RegulatoryInformation/Guidances/ucm073122.pdf

International Organization for Standardization. (2015). ISO 9000 Quality management (ISO 9001:2015). Retrieved from http://www.iso.org/iso/home/standards/management-standards/iso_9000.htm

Office for Human Research Protections. (2009). OHRP's compliance oversight procedures for evaluating institutions. Retrieved from http://www.hhs.gov/ohrp/compliance-and-reporting/evaluating-institutions/index.html

TransCelerate™ Biopharma Inc. (2013). *Position paper: Risk-based monitoring methodology* [position paper]. Retrieved from http://www.transceleratebiopharmainc.com/wp-content/uploads/2013/10/TransCelerate-RBM-Position-Paper-FINAL-30MAY2013.pdf

U.S. Food and Drug Administration. (2010). *Information sheet guidance: For IRBs, clinical investigators, and sponsors: FDA inspections of clinical investigators.* Retrieved from http://www.fda.gov/downloads/RegulatoryInformation/Guidances/UCM126553.pdf

7 Building Budgets

Clinical trials cannot be conducted without adequate funding. When considering a study, one of the first steps the study manager or appropriate budget staff must take is to determine if it is financially feasible for the site to conduct the study. Study budgets are complex and require analysis of the protocol, the clinical trial agreement (CTA), the informed consent, and an understanding of the regulations of the Affordable Healthcare Act that impact research costs. In addition, information on the cost of the study procedures, staff salaries including benefits, contractors, and facility and administration fees (overhead) is needed. This chapter covers the following budget steps: (1) identifying study costs to determine financial feasibility of the study, (2) budget development, (3) billing, (4) negotiating the budget, and (5) the payment plan.

FINANCIAL FEASIBILITY

Study feasibility was covered in Chapter 2 detailing why a site should or should not conduct a proposed study. In this chapter, we cover financial feasibility and whether the study site can afford to conduct a proposed study. The study team must look at all of the costs of doing a study, both direct and indirect costs, to answer this question. Compensation from the study sponsor should be adequate to cover all costs as well as bring in adequate revenue as defined by organization policy. Conducting a financial feasibility review enables the site to determine the cost per subject, the number of subjects required to break even, and the number needed to meet the revenue needs of the site.

Some costs are simply the cost of doing business and should be absorbed by the site. While sponsors are looking for ways to reduce or contain the costs of conducting studies, they are generally attempting to be fair to the study site. A sponsor, however, does not like being "nickeled-and-dimed" to death and it is important that the study manager understands the costs of doing business and does not include these in the study budget. Costs of doing business are those costs that the site would incur whether or not it was conducting the study. These include such things as general space, telephones, copy machines, waiting area, and business tasks such as signing financial disclosures, submitting Form FDA 1572, confidentiality agreements, and other documents.

Financial feasibility goes hand in hand with budget development. The steps covered in developing a budget in this chapter will assist the site manager in identifying costs of the study. Each of the steps presented in the development of the budget is an important part in the determination of financial feasibility and the site's decision of whether or not to conduct the study (see Figure 7.1 for budget development steps).

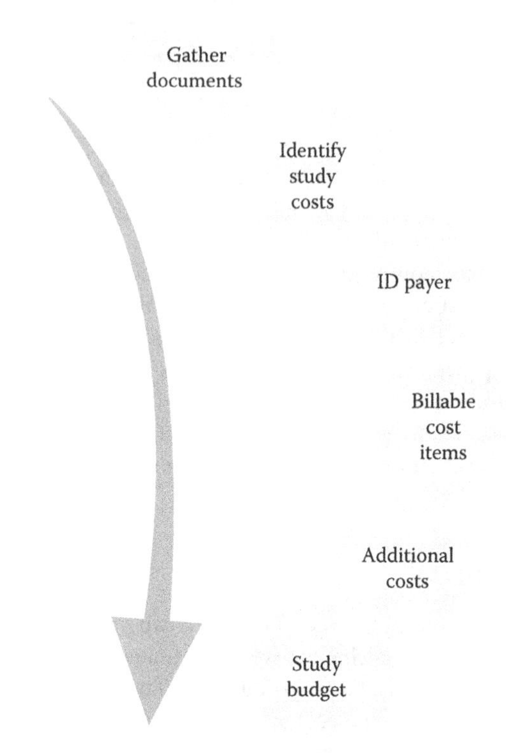

FIGURE 7.1 Budget development steps.

BUDGET DEVELOPMENT—INTERNAL BUDGET

The first step in creating a budget is to gather all of the study's essential documents, including the protocol, informed consent, lab and pharmacy manuals, sponsor budget, and the CTA. The information needed to create an accurate budget can be found in these documents. Do an initial review of the protocol and informed consent to become familiar with the study. As you review these documents, identify items that need clarification or questions. Examples may be type of monitoring and how often, any special reports required, lab clarifications, etc. Make sure these items are clarified before you sign or approve the study budget.

Study Procedures and Staff Time

Study procedures are identified through a thorough review of the protocol. Protocols include all procedures to be performed at each study visit as well as follow-up visits or situations that require additional testing or visits. Figure 7.2 is an example of a schedule of events (SE). This is similar to what will be found in a study protocol. The SE in Figure 7.2 will be used to develop a sample budget for this chapter. In addition

Procedure	Screen	Visit 1	Visit 2	Visit 3	Visit 4	Visit 5
Informed consent	•					
Vitals (weight, RPM, heart rate, blood pressure)	•	•	•	•	•	•
Height	•					
Medical history	•					
Physical exam	•					•
Adverse events		•	•	•	•	•
Concomitant medications	•	•	•	•	•	•
Blood draw	•	•	•	•	•	•
12-lead EKG	•					
Urine pregnancy test	•	•	•	•	•	•
Randomization		•				
Drug disbursement and collection		•	•	•	•	•

FIGURE 7.2 Schedule of events.

to the SE, it is important to review the descriptions of the visits and procedures in the protocol to identify costs/procedures that may not be included in the SE. A review of the lab and pharmacy manuals may be appropriate to identify any lab processes or procedures required on site, special handling, and others. Review all sections of the protocol for any discrepancies.

Tip: Use search and find—ctrl F—to look for labs, shipping, and other key words that are related to study costs when you are reviewing these documents.

From the SE, procedures for each visit can be identified. Once identified, apply a cost to each of the procedures. Institutions generally use either an institution charge master or a fee schedule to assign a cost to each procedure.

Charge Master

Refer to the institution's charge master for the current fees for each procedure. Of note here, the costs of procedures will vary from one geographic location to another. The cost to do a 12-lead electrocardiogram is likely to be more in San Francisco, California, than in Phoenix, Arizona. It is critical to make sure that the costs you apply are fair market value for the geographic location. Figure 7.3 is a mock-up example of what a portion of a charge master might look like, showing the costs and codes for venipunctures and magnetic resonance imaging of the spine. You will notice that for the procedures listed, there are multiple types (Figure 7.3). The type of institution will also factor into the cost; for example, medical center costs tend to be higher than those of a clinic. The protocol should specify the specific type of procedure for the study. If it is not clear, contact the sponsor to determine the specific procedure. When

1000000000	HB venipuncture blood draw	10.00
1000000001	HB research venipuncture blood draw	45.00
1000000002	HB exam evidentry blood draw	57.00
1000000003	HB capillary blood draw	10.00
1000000004	HB arterial puncture blood draw	95.00
1000000005	HB MRI C-spine wo contrast	3080.00
5000000001	HB MRI C-spine + contrast	3390.00
5000000002	HB MRI T-spine wo contrast	3140.00
5000000003	HB MRI T-spine + contrast	3390.00
5000000004	HB MRI L-spine wo contrast	2683.00
5000000005	HB MRI L-spine + contrast	3390.00
5000000006	HB MRI C-spine +/– contrast	5326.00
5000000007	HB MRI T-spine +/– contrast	5089.00
5000000008	HB MRI L-spine +/– contrast	5160.00

FIGURE 7.3 Sample charge master.

applying a cost for a procedure, it is important to make sure you are charging for the correct one. For example, a charge master may include 20 or more X-ray procedures, 40 or more blood tests, and so on. Review the visit details and what is included in each assessment/procedure to ensure that you have identified all costs for the procedure.

Cost of Study Visits

Once the cost of each procedure has been identified in the charge master, add the appropriate cost to the applicable procedure of the study visits (Figure 7.4). This is a start to building the budget.

Procedure	Screen	Visit 1	Visit 2	Visit 3	Visit 4	Visit 5
Informed consent	•					
Vitals (weight, RPM, heart rate, blood pressure)	•	•	•	•	•	•
Height	•					
Medical history	•					
Physical exam	•					•
Adverse events		•	•	•	•	•
Concomitant medications	•	•	•	•	•	•
Blood draw	$50.00	$50.00	$50.00	$50.00	$50.00	$50.00
12-lead EKG	$200.00	N/A	N/A	N/A	N/A	$200.00
Urine pregnancy test	$10.00	$10.00	$10.00	$10.00	$10.00	$10.00
Randomization		•				
Drug disbursement and collection		•	•	•	•	•

FIGURE 7.4 Study visit procedures.

The next step is to determine the costs of the remaining tasks/procedures, including the staff required to perform the procedure and the amount of time to complete each procedure. Note that some procedures may require more than one staff member. For example, a physical exam may be performed by a study nurse with physician oversight. The informed consent might be presented by the study coordinator with the investigator reviewing the study and answering questions in the last 15 minutes. Refer to the protocol to determine any specific requirements detailing who must perform procedures. In addition, the site might have policies or standard operating procedures that dictate who performs what procedures/activities in the conduct of clinical trials.

Staff time includes salary, benefits, and payroll taxes. Benefits might include the following: vacation, sick leave, medical and dental, retirement, life insurance, short-term and long-term disability, and eye care. Payroll taxes include local, state, federal, FICA (Federal Insurance Contributions Act [Social Security and Medicare]), and Workmen's Compensation. (For the budget presented in this chapter, a figure of 33% is applied to the hourly rate to account for both benefits and taxes.)

To estimate the time it will take staff to perform the various study procedures, review previous similar studies that the site has conducted. If the site has not conducted studies before or has not conducted a similar study, the site might contact other sites or use professional networks to gather information. Figure 7.5 breaks down the staff costs for each procedure into the following categories:

- Staff hourly rate;
- Hourly rate plus benefits and taxes (33%);
- Estimated time required for each staff member to perform the procedure;
- Staff costs based on amount of time spent; and
- Total staff cost of the procedure.

Once staff costs have been calculated, these costs can be added to the study visit procedures to determine total procedure cost (Figure 7.6). For a procedure fee identified in the charge master, staff costs must be added for a total cost. For example, the fee for the blood draw is $50.00. To determine total cost, the staff time must be added to the fee. (Blood draw fee = $50 + $13.30 staff cost, for a total cost of $63.30.)

Adding the costs of the procedures for a visit provides the total cost to the site to conduct that visit. For example, the total cost of the screening visit is $554.85 and the total cost for Visit 4 is $199.65. To determine the cost of all study visits per subject, simply add the total costs for each study visit (screen through Visit 5). The cost to the site for study visits is $1,853.00 per study subject. In addition to the study visit cost, to determine the total cost per subject for the study, subject compensation for time and travel needs to be added to the visit costs. If compensation is $25.00 per visit and there are five visits, add $125.00 to the visit costs, bringing the total cost to the site per subject to $1,978.00.

Procedure	Staff	Hourly rate	With benefits/taxes (33%)	Time	Staff costs	Total costs
Informed consent	SC PI	$40/hr. $200/hr.	$53.20/hr. $266.00/hr.	1 hr. 15min	$53.20 $66.50	$119.70
Vitals (weight, RPM, heart rate, blood pressure)	RA	$15/hr.	$19.95/hr.	30 min	$10.00**	$10.00
Height	RA	$15/hr.	$19.95/hr.	5 min	$2.40**	$2.40
Medical history	SC RN	$40/hr. $70/hr.	$53.20/hr. $93.10/hr.	45min 15min	$39.90 $23.30**	$63.20
Physical exam	SP	$150/hr.	$199.50/hr.	15min	$50**	$50.00
Adverse events	SC PI	$40/hr. $200/hr.	$53.20/hr. $266.00/hr.	30min 15min	$26.60 $66.50	$93.10
Concomitant medications	SC	$40/hr.	$53.20/hr.	15min	$13.30	$13.30
Blood draw	SC	$40/hr.	$53.20/hr.	15min	$13.30	$13.30
12-lead EKG	SC RA	$40/hr. $15/hr.	$53.20/hr. $19.95/hr.	30 min 30 min	$26.60 $10.00**	$36.60
Pregnancy test	LT	$20/hr.	$26.60/hr.	15min	$6.65	$6.65
Randomization	SC	$40/hr.	$53.20/hr.	15 min	$13.30	$13.30
Drug disbursement and collection	SC	$40/hr.	$53.20/hr.	15min	$13.30	$13.30

** Numbers rounded

Study coordinator = SC Research nurse = RN Investigator = PI
Study physician = SP Lab Tech = LT Research assistant = RA Data entry = DE

FIGURE 7.5 Study procedure—staff costs.

The sponsor's proposed budget, sometimes referred to as the external budget, generally does not break down the costs to the extent required by the site to determine financial feasibility. Sponsor budgets typically propose compensation to the study site in one of three ways:

1. Proposed compensation per study visit;
2. Proposed compensation per subject; or
3. Propose compensation for total subjects to be recruited.

Conducting the financial feasibility allows the site to determine if the sponsor's proposed compensation, regardless of which way it is presented, provides

Procedure	Screen	Visit 1	Visit 2	Visit 3	Visit 4	Visit 5
Informed consent	109.40					
Vitals (weight, RPM, heart rate, blood pressure)	10.00	10.00	10.00	10.00	10.00	10.00
Height	2.40					
Medical history	63.20					
Physical exam	50.00					50.00
Adverse events		93.10	92.10	93.10	93.10	93.10
Concomitant medications	13.30	13.30	13.30	13.30	13.30	13.30
Blood draw	63.30	63.30	63.30	63.30	63.30	63.30
12-lead EKG	236.60					236.60
Urine pregnancy test	6.65	6.65	6.65	6.65	6.65	6.65
Randomization		13.30				
Drug disbursement and collection		13.30	13.30	13.30	13.30	13.30
Total visit cost	**554.85**	**212.95**	**199.65**	**199.65**	**199.65**	**486.25**

FIGURE 7.6 Total procedure cost.

Type	Cost to site
Compensation per study visit Screening Visits 1 Visits 2–4 ($199.65/visit) Visit 5	 $554.85 $212.95 $598.95 $486.25
Compensation per subject (all visits plus $125 subject honorarium)	$1978.00
Compensation for total subjects (eight subjects)	$15,824.00

FIGURE 7.7 Study subject costs.

compensation for the costs to the site. Figure 7.7 shows the costs to the site in the three ways typically presented by a sponsor.

Other Study Costs

Visit costs are only one part of the cost of doing the study. There are many other costs associated with conducting a study.

Reviewing the SE is a good place to start. However, it is important to review the entire protocol for hidden items and services. Items such as subject compensation; extra services for certain lab results, which could include follow-up tests; and local labs as opposed to central labs are generally not found in the SE. Reviewing the inclusion and exclusion criteria is important. Some criteria may require additional testing that does not show up in the SE, for example, HIV testing or review of rheumatoid arthritis factors. Pay attention to footnotes. Does the footnote affect budgeting and billing? For example, a procedure may include timing the results at multiple intervals, such as during a five-hour intravenous infusion requiring a nurse to check vitals every 30 minutes, increasing staff time.

Some study costs are paid by the sponsor upon receipt of an invoice from the study site. These are costs that do not occur on a regular basis and are typically referred to as "pass-through" costs. Many of these costs may or may not occur during the conduct of the study. Examples of costs to be invoiced might include items such as subcontractor fees, record storage, subject travel and parking, subject meals for lengthy visits, screen failures, adjudication of serious adverse events (SAEs), and shipping and handling of biological specimens. These items are generally invoiced to the sponsor when they occur. Although these costs might not occur, a set fee should be determined and included in the budget as an invoiced item when possible. For example, the cost of adjudicating an SAE might be set at $500 per SAE and screen failures would be set at the cost of the screening visit, which is $547.55.

Additional costs to consider might include items such as data entry, management/data verification/queries, institutional review board (IRB) amendments and reports, reconsenting, room fees (cost—for example, patient in room 8 hours; hospital overnight stays and travel and lodging, for subjects who travel a long distance, etc.), remote monitoring fee (coordinator time to prepare records, be available, follow-up, etc.), and additional testing/procedures for adverse events. Only a thorough review of the protocol will assist in the identification of all of the costs associated with the conduct of the study.

Study Start-Up Costs

Preparing the site to conduct the study involves costs to the site. These are costs that the site would not incur if not doing the study. These costs should be covered by the sponsor as a one-time nonrefundable fee. These costs include IRB fees for the submission and approval of the informed consent and other study documents and administrative fees for budget and contract review and negotiation, protocol review, study-specific training, investigator meetings, site set-up, chart reviews for recruitment, and source documentation preparation. Depending on the type of study, there may also be specialty fees such as special equipment needed, set-up fees, radiology, pharmacy, laboratory, pathology, etc. Once the site has determined the costs to start the study, this should be included in the budget as a one-time nonrefundable fee. This cost should be nonrefundable because, whether or not the site is able to enroll subjects into the study, the site still incurs the costs.

Recruitment

Recruiting subjects, whether doing so through chart reviews, social media, or other advertising avenues, is also a study expense. Most sponsors provide recruiting materials and a set amount for recruiting. Site personnel need to determine whether the sponsor's proposed budget will cover the recruitment activities needed to recruit the number of study subjects to meet the site's recruitment goal.

Shipping and Handling

In multisite clinical studies, sponsors may contract with a central lab to process samples collected from the subject. Shipping biologic materials requires special handling as well as training. Shipping and handling costs should be covered by the study sponsor and included in the budget.

SAE Resolution

When an SAE occurs, the site is responsible for following the event through resolution. This includes obtaining all necessary records from all parties who were involved in the SAE, which might include physician records, hospital records, and death certificates. Following the event to resolution can take months and sometimes even years until the final outcome is known. Tracking and obtaining records are onerous tasks and take perseverance and time. Institutions generally charge to copy patient records. It is not possible to foresee SAEs or the time and effort it will take to resolve the SAE. These may be added to the sponsor's proposed budget but often are not. This might be a cost the site negotiates as a flat fee per SAE or for actual costs of adjudicating the SAE. To determine an equitable flat fee, peruse similar studies and determine the cost (including staff time, copies of records, and shipping costs) of resolving SAEs in that study. These costs can be used as a starting point to determine and negotiate reimbursement to the site for SAE adjudication.

Subject Drop-Out

When a subject drops out of a study, costs are incurred by the site to ensure the safety and welfare of the subject. The best scenario would be that subjects inform the study site coordinator when they no longer wanted to participate in the study. In this case, a final study visit can be scheduled with the subject. At this visit, it is important to address any safety issues and to obtain final vital statistics and pertinent data. This would be considered an unscheduled visit and included as an invoiced cost billed to the sponsor based on the time and procedures completed at the visit.

Unfortunately, what usually occurs is that a subject stops coming to the visits without notice. When this occurs, it is the responsibility of the site to contact the subjects and bring them to the site for one final safety visit. This can take many attempts and time by the coordinator. Appropriate attempts must be made and documented if the site is not able to reach the subjects or they refuse to come in for a final visit.

Monitor Visits

Monitor visits require site staff time when they occur. Risk-based monitoring may include remote and on-site visits. Both types of visits require staff time. Staff need to prepare for either type of monitoring by having records ready and accessible as well as being available during monitoring to answer questions and resolve queries. Monitors generally request a meeting with the study investigator at the end of the visit. Time spent by staff during monitor visits should be included in the budget. The sponsor's monitoring plan should be detailed in the CTA. Reviewing these details will provide the site with an idea of the staff time required for a monitor visit and assist in determining the cost to the site of the monitor's visit.

Early Termination of a Study

When a study terminates early, the site continues to incur costs. The site must contact the study subjects and bring them in for a final visit. The site must enter all data and resolve any existing queries. The site must close out the study, which includes drug/product accountability, return of product to the sponsor, completing subject records, and storage of study records. The CTA should clearly define the processes to be carried out if a study is terminated early. Review the CTA to make sure that the sponsor will cover early termination costs to the site and that there are provisions in the budget to charge these costs to sponsor.

Audits/Inspections

When an audit by the sponsor or inspections by the Food and Drug Administration (FDA) occur, costs are incurred by the site. Audits or inspections might occur for several reasons as discussed in Chapter 6. Often when a drug/device is being reviewed by the FDA for approval to market, the FDA will conduct inspections of study sites who were involved in conducting the trials. An inspection might be for cause, where the FDA has received a complaint about the site or the sponsor and is following up on that complaint. Sponsors sometimes conduct their own audits to review not only the site's performance but also that of their monitors. These audits/inspections require prep time, staff time required during the audit or inspection, and copying records. If you have been through an audit or inspection before, you will have a general idea of the time and effort required and can estimate a reasonable charge.

Study Close-Out

Upon completion of the study and after data lock (the point where data from all sites have been verified and all queries resolved), the sponsor will schedule an appointment for a study close-out visit. During this visit, final drug accountability will be completed by the site and monitor. Generally, remaining investigational product will be returned to the sponsor. The monitor will also review record storage and any other expectations that the sponsor may have. Shipping costs plus staff time for this visit should be covered by the sponsor.

Indirect Costs

Indirect costs are the costs that are not directly related to the study but are incurred by the site to conduct the study. These are also known as overhead or facilities and administration costs. These costs include administrative support, phones, faxes, general supplies, and space. Generally, these are figured at 25%–35% of the total budget but may be higher or lower based on the site. Indirect costs are typically figured once all of the other study costs have been determined. Applying the indirect cost will provide the total budget costs to the site. For example, if the total budget is $525,000 and overhead costs are 28%, applying the 28% overhead brings the total cost to $672,000 for the site budget.

Hidden Costs

It is important to identify any additional costs incurred by the site to the conduct of the study. Some of these may include biohazard waste disposal and destruction of confidential records. When a site is involved in a multicenter study, the investigator is required to review the MedWatch reports of serious events from all sites. This can involve a large number of reports and time to review. Although reviewing these reports may be considered a cost of doing business, this should not require an excessive amount of the investigator's time. If the study is complex, the number of reports that the investigator must review may be excessive. In this event, it might be reasonable to negotiate with the sponsor to compensate for a certain number of reviews.

Cost of Living

Many phase 3 studies can last for years. If this is the case, the site budget should include cost of living increases each year.

A sample of a complete study site budget is found in Figure 7.8.

Budget costs can be categorized as shown in Figure 7.9.

BILLING AND COVERAGE ANALYSIS

After determining the costs for the study, the next step is to conduct a coverage analysis. The coverage analysis identifies which parties will reimburse the site for the various study costs. If the sponsor is covering all of the research costs, this step can be eliminated. For coverage analysis, first, identify the items the sponsor is agreeing to pay. Refer to the sponsor's proposed budget to identify these items. Mark these items on your budget as research items (sponsor paid).

Once all of the research items have been marked, the next step is determining what party will cover the costs of the remaining items. According to the Affordable Care Act ("US Department of Health and Human Services," 2010), insurers may not limit coverage for items that would otherwise be provided if the patient was not enrolled in the study. Many standard of care (SOC) items and services can be billed to the patient's insurance. Standard of care are those procedures or services that would be normally covered by the subject's insurance, such as HgA_{1C} tests for persons with diabetes. Of note, reimbursement for items will be based on the individual's coverage; therefore, there should be a method included in the budget to invoice the sponsor if an SOC is not covered by the individual's insurance plan.

Cost item	Per unit	# Units	Total
Study start-up			$10,000.00
Investigator meeting			
Investigator			$1200.00
Study coordinator			$600.00
Study visits			
Screen	$547.55	8	$4380.40
V1	$231.95	8	$1855.60
V2	$231.95	8	$1855.60
V3	$231.95	8	$1855.60
V4	$231.95	8	$1855.60
V5	$505.25	8	$4042.00
Subject compensation	$25.00	5	$125.00
Monitor visits	$1200.00	4	$4800.00
Additional staff			
Recruiter	$45,000.00	0.25	$11,250.00
Receptionist	$30,000.00	0.1	$3000.00
Billing	$400.00	12	$4800.00
Data entry	$32,000.00	0.5	$16,000.00
Recruitment			$12,000.00
Subtotal			**$79,619.80**
Site overhead (30%)			$23,885.94
Total fixed costs			**$103,505.74**
Invoiced items (billed/occurrence)			
IRB continuing review reports	$125.00	TBD	
Screen failure	$547.55	TBD	
Subject re-consent	$40.00	TBD	
SAE adjudication	$500.00	TBD	
Shipping and handling	TBD	TBD	
Review of MedWatch reports in excess of 200	$50.00	TBD	
Unscheduled visits (cost determined by procedures performed)	TBD	TBD	
Record storage	TBD	TBD	
Sharps containers	$12.00	TBD	
Shredding of confidential documents	$80.00	TBD	

FIGURE 7.8 Example study site budget.

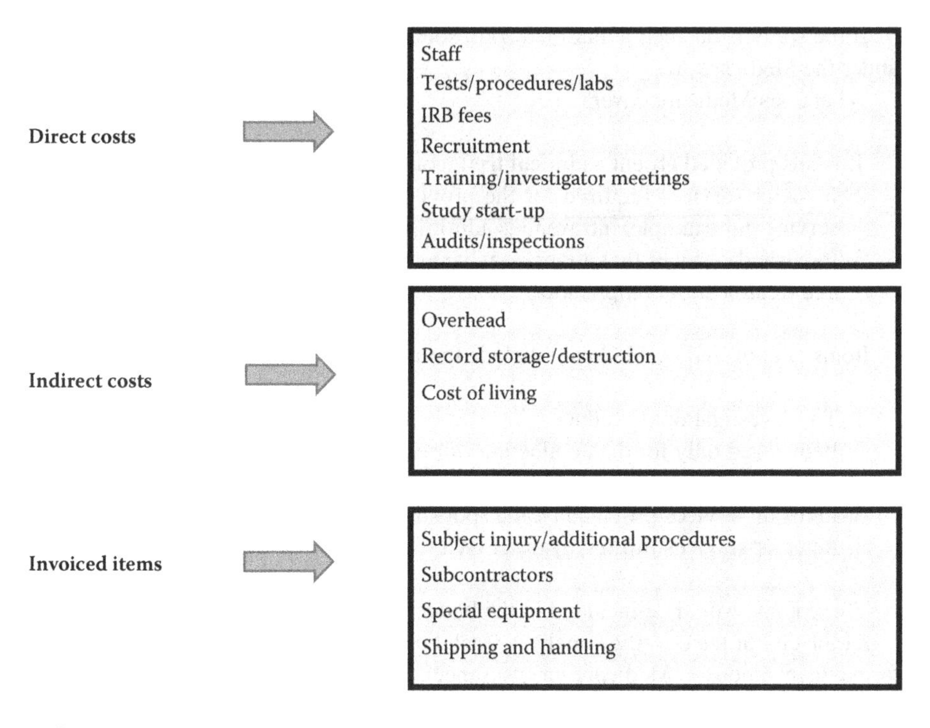

FIGURE 7.9 Budget cost categories.

If the study meets the qualifying requirements under the Affordable Care Act, some items may be covered by Medicare under National Coverage Determination (NCD) or Local Coverage Determination (LCD). Medicare's Clinical Trial Policy ("Center for Medicare and Medicaid Services, 2007," NCD 310.1) addresses the items and services that qualify for Medicare coverage within a clinical trial. National Coverage Determinations can be found on the Center for Medicare and Medicaid Services website. Some states also have LCDs, which may differ from NCDs. If a state has LCDs, these fees would apply as opposed to NCD fees.

Medicare rates include Clinical Lab and Physician Fee Schedules. The fees can be found at cms.gov/Medicare under "Fee for Service Payment" under "Schedules." Fees are listed by Current Procedural Terminology codes and by state.

For a clinical trial to qualify under NCD310.1, it must meet the following four criteria:

1. The investigational item must fall under a Medicare benefit category (Drugs and Biologics, Transplant Services, Inpatient Hospital Services). The trial must be designed to test efficacy, not just safety.
2. The trial must have a therapeutic intent.
3. The trial must enroll patients with the diagnosed disease.
4. The study must be deemed (to qualify for Medicare coverage) (having an Investigational New Drug, cooperative group, funded, etc.).

If the study is deemed a qualified trial, some of the SOC costs might be covered under the Medicare Act.

What does Medicare cover?

1. Items provided absent a clinical trial (conventional care)
2. Items or services required for the provision of the investigational item or service (for example, intravenous administration of an investigational drug)
3. Items and services that are reasonable and necessary care for the diagnosis and treatment of complications

Items not covered by Medicare include the following:

1. The investigational product
2. Items used only for data collection (for example, multiple scans during the trial for data collection or multiple blood draws for data collection)
3. Items or services provided by the sponsor free of charge
4. Items or services limited by other NCDs or LCDs

These items would be included in the budget as research (sponsor paid) items.

At the end of the coverage analysis, each budget item will have a party identified for payment (sponsor, Medicare, or insurance). You will need to establish a payment plan for each of the parties, ensuring that no item is double charged.

EXTERNAL BUDGET

Generally, the sponsor provides a proposed budget template to each site. This is referred to as the external budget. The external budget includes all of the study costs that will be compensated by the sponsor. Use the budget template provided by the sponsor as a starting point. Review and compare the costs submitted by the sponsor to the costs that you determined during the creation of the internal budget. If a fee is less on the sponsor's proposed budget, transfer the internal budget figure into the sponsor's budget template. Add all costs to this budget including, direct, indirect, items to be invoiced, and hidden costs. You may need to add line items to the budget.

This is where the negotiation begins. Thorough review of the protocol and other study documents, completing all steps of budget development, and conducting a coverage analysis should provide justification for the costs that you are requesting.

BUDGET NEGOTIATION

Negotiation is about communication. Being prepared is critical to successful negotiation. You will need to provide justification to support requested fees. Data you collected to create your budget will provide support for your requests. Historical cost data from your site on the costs of similar studies can be used as additional data to justify requests.

Realize that negotiation is a two-way street. Both parties are attempting to negotiate a budget that is most beneficial to them. Do not go into the negotiation process with your bottom-line figures. Bottom line is the minimum amount determined by the site to properly conduct the study. Using the bottom-line figure leaves no cushion to work with for further negotiation. Negotiation involves give and take and generally requires multiple rounds before an agreement is reached. Most sponsors have experience conducting clinical trials and are generally familiar with the costs and attempt to be fair to the study site. The sponsor will use these general costs to provide the basis of their budget proposal.

Parke (2013) proposes several guiding principles for sponsors when negotiating budgets and contracts. These same principles can be applied to research sites.

- Negotiate strategically including choosing an appropriate atmosphere for meetings and calls. It is to your benefit to set a time as opposed to discussing terms off-the-cuff. During the process, write down questions you have and make sure that these are answered. You might be able to use the answers to strengthen your requests during the negotiation process. Think about standards that may support or provide leverage for your requests. Decide what issues to discuss first and control the agenda.
- Be prepared. The more you prepare for the negotiation, the better you will do. Negotiations can be quite complex and preparation is a must. Make sure you are very familiar with the protocol, the budget, and the CTA. Clarify terms and definitions in the budget and contract. For example, how is a screen failure defined or what is defined as enrolled in the study?
- Establish a reputation during the negotiations as being honest and trustworthy. Deal with the sponsor in a straightforward and professional manner. Do not delay feedback during the process.
- Gather information throughout the process. If you make commitments, ensure that you keep them. "Know when to stop talking and listen" (Parke, 2013, p. 43). Communication is the key to success.
- Maximize your leverage. Use enrollment history, success in conducting trials, meeting deadlines, and providing quality data as leverage points during the negotiation.

If there is an item you do not understand, ask questions. Make sure that all issues and questions are clarified to your comfort. Understand what you are agreeing to.

Realize that it does not hurt to ask. Prioritize your needs. Define what you need to be able to conduct the study and items that you could do without. The final outcome of the negotiation process may be to decline the study. Agreeing to lower rates can set your site up for failure. This also can create the impression that your site will do future studies at a lower rate.

Some items to consider in your negotiations include the following:

- Screen failures—avoid a cap. Ensure the sponsor that the site will make the best effort to enroll subjects and the site needs to be paid for services rendered. Look at inclusion and exclusion criteria when seeking reimbursement for screen failures. The number and ranges for inclusion and exclusion criteria may impact the number of screen failures. Some studies are much more difficult to recruit subjects.

- Pediatric studies are usually more time-consuming. You may spend more time dealing with parents or guardians, who have many questions, than with the subject.
- Complexity of the study impacts not only the recruitment but also investigator and study staff time.
- As mentioned earlier in this chapter, the geographic location of the site will impact the costs of services and procedures.

BILLING PLAN

Understand what triggers payments from the sponsor. The payment plan is included in the CTA, which is presented in Chapter 8. Site policies may allow only certain people access to the CTA, but at minimum, you will need a copy of the payment plan portion of the CTA to understand the billing plan and how the site will be paid.

There are various payment types. These are the three most common:

- Per visit—payments are made per subject visit completed. Make sure that you know what is expected. For example, does payment require approval/verification by the monitor or sponsor of the Case Report Form?
- Milestone—payments are made based on completion of preset milestones. Examples of milestones might be payment upon five patients screened; at completion of the third, fifth, and ninth visit, and so on. Be aware that if a patient completes some visits but does not reach the next milestone, the site may not be reimbursed for visits conducted.
- Invoiced items—payments are made upon the sponsor's receipt of invoices from the site. Consider timing of payment to ensure that it is a reasonable turnaround time for payment of invoice. Does invoicing involve an excessive amount of staff time or use of an accountant? If so, consider adding a processing fee to the budget.

Carefully review the sponsor's payment plan. Questions that the site representative should ask include the following: Are the terms reasonable? When can the site expect the first payment? Do the terms meet the needs of the site? Does the payment frequency/timing (monthly vs. quarterly) cover site costs such as payroll? Be cognizant of long intervals in between payments as these can have a major negative impact on the site's cash flow.

If the payment plan includes site invoicing, make sure that it is clear what must be invoiced and how often. Know the deadlines for submitting invoices and time to expect receipt of payment. Develop an internal system to track invoices and payments.

Final payment—know when the final payment will be made to the site. Make sure it is based on your site specifically, not on the entire study. Often, a sponsor may retain a certain amount or percent of the budget as "hold back." If this is done, make sure the payment plan clearly identifies when the site will receive the "hold back" payment (for example, within 60 days of final data lock for the site, within 30 days of the close-out visit, etc.).

A VIEW FROM INDIA

The process of calculating and validating the internal budget, other study costs, external budget, budget negotiation, and the billing plan is quite similar in India as in the United States.

As mentioned in this chapter, site personnel must consider all the costs when compiling an internal budget, which includes both direct and indirect costs, invoicables, and other costs to determine the financial feasibility of a clinical study. The sponsor's proposed budget, which is the external budget, generally does not break down the costs to the extent required by the site to determine financial feasibility; nonetheless, the site should be able to determine if it is practical.

Budget terminology is slightly different in India. For example, site budgets usually consist of a fixed and a variable component. The fixed component provides compensation (salary) for the site coordinator and use of the site infrastructure such as telephone, fax machines, printer, refrigerator, or storage cabinet. The variable component, which is protocol dependent, includes investigator fees and is calculated per completed subject (Varawalla & Jain, 2011).

The site must ensure that the sponsor will provide the SOC for each study subject. Compensation of the investigator and study staff time is to be provided by the sponsor, as they are not covered by insurance.

SAE Resolution

There is a noteworthy difference in the way compensation for SAEs is handled in India as compared with the United States. Because there is no equivalent of an Affordable Care Act in India, site personnel must check to ensure that the sponsor provides insurance for all the study subjects. As per regulations, and as mentioned in the good clinical practice guidelines published by Central Drugs Standard Control Organization (CDSCO), the primary responsibility is that of the sponsor (a pharmaceutical company, a government, or an institution) to provide compensation for any serious physical or mental injury arising out of a clinical trial for which subjects are entitled or to provide insurance coverage for any unexpected injury (Government of India, Ministry of Health & Family Welfare, CDSCO, 2004). It is important that site personnel check that the sponsor complies with the latest guidelines for compensation related to SAEs that result in the injury or death of subject. Both the guidelines and the formulas for compensation can be found on the CDSCO website.

CHAPTER REVIEW

In this chapter, we reviewed the steps of budget development to determine whether or not it is financially feasible for a site to conduct a study. The steps involved in creating a budget include gathering and reviewing study documents, identifying costs from the different documents, figuring staff time, and doing a coverage analysis.

We reviewed the keys to successful budget negotiation and some of the costs to address in negotiations with the sponsor. We also reviewed payment plans and the common payment types.

APPLY YOUR KNOWLEDGE

1. You tasked a study manager with reviewing the study protocol to determine the cost to the site to conduct the study. After review, the manager presents a budget based on the SEs (calculating staff time and procedure costs per visit). In addition, the budget includes indirect costs, subject compensation, IRB fees, recruitment costs, and items to be invoiced. You have to determine if it is feasible for the site to conduct the study. Do you have adequate information to make this decision? Yes or no? Explain.
2. During budget negotiations, your site has requested payment for the following items from the sponsor:
   a. A nonrefundable start-up fee of $7,000 for site preparation
   b. A fee of $500 for the adjudication of each SAE
   c. Cost of study coordinator time during site monitoring visits

   The sponsor has agreed to the adjudication fee for SAEs. However, the sponsor team believes that the start-up cost and the coordinator time during monitoring visits should be considered a cost of doing business by the site. The site's position is that these costs would not be incurred by the site unless they were conducting the study for the sponsor, and therefore, they are costs of doing the research, not costs of doing business. What types of information would you present to justify these as costs of doing research to the sponsor during the next negotiation meeting?

REFERENCES

Center for Medicare Services. (2007). National Coverage Determination (NCD) for routine costs in clinical trials. NCD 310.1. Retrieved from https://www.cms.gov/medicare-coverage-database/

Government of India, Ministry of Health & Family Welfare, Central Drugs Standard Control Organization. (2004). Good clinical practice for clinical research in India. Retrieved from http://www.cdsco.nic.in/html/GCP1.html

Parke, J. (2013). Negotiating effective clinical trial agreements and study budgets with research sites. *Applied Clinical Trials*, 22(4):40–44.

U.S. Department of Health and Human Services. (2010). Affordable Care Act. Retrieved from www.hhs.gov/opa/affordable-care-act/

Varawalla, N., & Jain, R. (2011). Clinical trials in India. In Chin, R., & Bairu, M. (Eds.). *Global clinical trials: Effective implementation and management* (pp. 119–157). Elsevier, Amsterdam.

8 Contracts, Clauses, and Closing the Deal

The clinical trial agreement (CTA) is one of several documents that govern the conduct of human subject research (Pfeiffer, 2014). Food and Drug Administration (FDA)-regulated medical product trials sponsored by pharmaceutical companies require a CTA. The CTA is a legal and binding contract that governs the conduct and responsibilities of the parties involved in the clinical research study (Webb, 2006). The CTA is the foundation on which the relationship between the sponsor, site, and the investigator is built (Pfeiffer, 2014). Often, CTAs are difficult to understand because of the legal verbiage. Understanding the CTA is critical for study staff as a simple human error may cause a breach of contract and result in serious financial and legal implications for a study site. This chapter will provide an overview of the regulations related to CTAs, key sections of the CTA, contract language, understanding the contract provisions, and negotiating contracts.

In the United States, contracts fall under the jurisdiction of state law. Even so, because of their use in human subject research, there are a number of federal laws that are applicable to the CTA. Although these laws do not cover requirements of the content of the CTA they must be considered and addressed in the CTA (U.S. Department of Health and Human Services, FDA, 2010).

The FDA simply states that CTAs must be in compliance with laws and regulations applicable to human subject research. Examples of clauses in U.S. Federal regulations applicable to human study research can be found in Appendix A in the table titled "Federal Regulations Governing Human Research."

It is important to understand the language and terms of the contract as they may be defined differently than how the term or word is used in everyday language. Pay attention to the contract language, especially terms that you do not understand.

- Look for terms or words that are capitalized throughout the contract. They are capitalized for a reason. Make sure these terms are defined and you understand them.
- Be careful of broad terms such as "any" or "all," "immediately," "without limitation," and "best effort." These terms indicate that the site will drop everything and do whatever is needed. Replace these words with "reasonable efforts," "promptly," and the specifics for terms such as "any" and "all."

TERMS AND PROVISIONS OF A CTA

There are many clauses included in a CTA. The key clauses/terms of the CTA are listed and described in the following sections of this chapter.

Parties to the Contract

A contract has at least two parties. The party that is requesting a service and the party that is able to provide the requested service. In a CTA, there are often more than two parties involved in a contract, including the sponsor, the research site, and the investigator. Each party to the contract must be properly and formally identified with names, addresses, and contact information.

Scope of Work

The "Scope of Work" defines in broad terms the duties and deliverables of the contract parties. Tasks assigned to the investigator are defined in a number of areas in this provision, including his or her role in respect to the supervision of the study, obtaining informed consents from study subjects, delegation of authority, management of protocol amendments, recruitment and enrollment, and handling of human materials if applicable (Model Agreements and Guidelines International [MAGI], 2016). In addition, the general responsibilities of the sponsor such as reporting to the FDA and monitoring data are also included in this section.

Principal Investigator

This section defines the relationship of the investigator to the research site conducting the study. Typically, the investigator is either an employee of the research site or a contractor. This section also defines the investigator's responsibility for oversight of the study and outlines the process to be taken if the investigator resigns or is no longer able to conduct the trial.

The sponsor has many expectations of the site in conducting a clinical trial. These may be outlined in this section or there may be a separate section covering site deliverables. Deliverables include the following:

- Recruitment of subjects
- Obtaining subject consent prior to conducting any study procedures
- Completion of all tasks/procedures in the study protocol
- Collection and management of data
- Reporting
- Handling and storing of investigational product
- Handling and shipping of labs

Sponsor Obligations

Sponsor obligations include all services, study supplies, investigational product, and equipment that is provided by the sponsor. This section also outlines the sponsor's responsibilities to the FDA, including submitting the investigational product application, providing requested documents, monitoring the study, and reporting. While it is the site/investigator's responsibility for institutional review board (IRB) oversight and approval, the sponsor may elect to use a central IRB, relieving the site of this responsibility. In this circumstance, the site must obtain the site-specific IRB approval and submit continuing review reports for the site. The sponsor also has duties to the study subject. This is referred to as a fiduciary responsibility. The fiduciary relationship between the sponsor and the study subject should be detailed in this section. A fiduciary relationship creates two legal duties for the sponsor: (1) a duty of care and (2) a duty of loyalty. The duty of care is addressed in the informed consent process. The duty of care is to protect the safety and welfare of the study subjects. The consent outlines the risks and benefits of the study as well as the risks of the study procedures. A duty of loyalty implies that the sponsor will act in the best interest of the subjects (MAGI, 2016).

Reports, Reporting, and Audits

Outlined in this section are the responsibilities of the investigator and the site with respect to the management of study records. It clarifies who has access to the study documents and what level of access is permitted for the various study roles. Also included here is the management of regulatory audits, including the sponsor's expectations of the site. Ownership of the study data and records is defined in this section.

Monitoring Plan

The sponsor's monitoring plan should be clearly outlined in this section. Details should include how often the sponsor will monitor the site and expectations, including number of days the monitor will be at the site, type of space and equipment needed while on site, expectations of site staff time (coordinator, investigator, etc.), and items that will need to be accessed (patient records, pharmacy, etc.) and reviewed (all case report forms [CRFs] and source documents vs. a percentage). This section should have verbiage that includes how much advance notice the site will receive and that the monitor visits will be conducted during the site's normal business hours.

Sponsors may use risk-based monitoring, which was defined in Chapter 6. Risk-based monitoring that is conducted centrally (off-site) may require more administrative support and may require an additional time commitment from study staff. It is important to have the expectations defined so both the sponsor and the research team have clear expectations. Additionally, risk-based monitoring, although not done on site, should still be scheduled with the site.

Payment Plan

The payment plan was discussed in Chapter 7 in relation to study budgets. However, payment plans are often a point of contention between investigator/sites and sponsors and it is important to reiterate it here too. Having a detailed payment plan that outlines how the site will be paid, triggers for payments, and deliverables required to receive payment is critical.

Payment plans should include the following:

- Payment types—How will the site be paid? By visit, by milestone, or by invoice as outlined in Chapter 7.

- Frequency—How often will payments occur? Monthly, quarterly, or by milestone.
- Triggers—What triggers a payment? Triggers may include the following:
    - CRF submission
    - Data monitored
    - Number of patients enrolled
    - Completed patient visits

Confidentiality and Confidential Materials

In this section, it is to the benefit of the site that language be specific as opposed to generic. Specific language will assist the site in knowing what materials are considered confidential. Generic language might leave this open to interpretation and result in a site staff member divulging confidential information not understanding that it was confidential.

To prevent the accidental disclosure of confidential information, all sponsor confidential materials should be clearly marked "CONFIDENTIAL" on each page of a document. In addition, confidential information shared orally should be in writing and marked "CONFIDENTIAL" within a set time frame (for example, 3–5 days). The process for this and timeline need to be spelled out by the sponsor. For example, a monitor may share confidential information about the preliminary results of the study or the chemical structure of the investigational drug.

This section will also include the sponsor's requirement that study staff sign confidentiality agreements. These agreements protect the sponsor from the investigator and site personnel sharing trade secrets or disclosing study information to individuals who have not signed a confidentiality agreement.

Data Access

Defined in this section is who has access to the study data and when. Study data is owned by the sponsor. However, for scientific integrity and the dissemination of scientific knowledge it is important that the investigator has access to the site's data to present at scientific forums or for publications. For multi-center trials this section should define when the site/investigator will have access to the data from the entire study and how this access will be granted.

Liability and Subject Injury

This section defines responsibility for subject harm, injury, or death resulting from participating in the study and the costs that are covered. All medical and other injury-related expenses should be covered by the sponsor if the injury is a result of a subject's participation in the study.

Publication

Many investigators intend to publish the results of the studies at their sites as a way to advance scientific knowledge. Language in this section should give the investigator the right to publish or share study data in scientific forums and publications. The sponsor should have the right to review, but only to remove any confidential information. It is important to define the period of review for the sponsor (for example, 3–6 months) so that publication or presentations are not delayed indefinitely.

Indemnification

Indemnification protects one party from being liable for the negligent acts of the other party. This section should clearly state that the investigator and research staff are not liable for injury or harm caused by the negligent acts of the sponsor. Conversely, the sponsor is not liable for harm or injury caused by negligence of the investigator or site study staff. In addition, the investigator and site should be indemnified as long as they are conducting the study according to the protocol.

Jurisdiction

Contract law is governed by state laws. Because each state has different laws, it is important to know which state will have jurisdiction over the contract. If there are issues with the contract, resolution will occur in the state of jurisdiction, whether in a state court or by arbitration. Depending on where the issue will be resolved this could be costly to the site if it involves travel and time from work.

Contract Term and Termination

The contract term, the length of the contract, and termination of the contract should be clearly identified. Contract term includes the start date and end date of the contract. Oftentimes, trials run for a longer time than anticipated, so the procedure to extend the end date should be defined.

It should be spelled out in detail who can terminate the contract and the process to terminate. Included should be the length of notice required and type of notice to the other party from the party wanting to terminate the contract.

UNDERSTANDING THE CTA

Understanding the CTA allows an investigator/site to ensure that the contract is fair and balanced, to protect the investigator/site, and the research subjects. Not understanding the terms of the CTA can lead to significant legal and financial issues for the investigator/site—a simple misunderstanding might result in a breach of contract, or an issue regarding liability.

It is important that all parties to the contract have a clear understanding of what they are agreeing to and how compensation will be determined (Windschiegl, 2015). Reviewing the contract is the first step to understanding the contract. Any terms or language that the contracting parties do not understand should be clarified. It is not unusual for contracting parties to request a definition of terms included in the contract.

The contract should also be reviewed to determine that it is in compliance with current applicable laws and regulations. Generally, there is a statement in the CTA that affirms that the sponsor and the investigator/site will be in compliance with all current applicable regulations.

SITE-SPONSOR ISSUES

According to Pfeiffer, in a 2013 survey of U.S. investigators, respondents reported that budgets and payments are the most common issues they have with sponsors (Pfeiffer, 2014). This supports previous surveys of clinical research site administrators who also reported these as a common issue.

In addition to these, other issues reported include publication of study results, access to study data, indemnification, and subject injury language. Each of these terms has been presented in this chapter, including what needs to be included in the contract sections to protect the site. Budgets and payments were covered in Chapter 7. Thoroughly reviewing the contract for content will assist in minimizing issues with the sponsor related to the CTA.

NEGOTIATION

In Chapter 7, we covered negotiation items for the budget. Other than revising contract language so that terms are defined and understandable, the items covered in budget negotiation also apply to the CTA.

Most people are not born negotiators. Many individuals are uncomfortable in negotiation situations. However, it is a skill that can be learned through practice and experience. The most important factor in successful negotiation is being prepared by understanding the terms and definitions of the contract. Additionally, a successful negotiation includes knowing what is being agreed upon. Having documentation to justify requests, as discussed in budget negotiation in Chapter 7, will assist the negotiation. Documentation includes not only costs to do the trial but also justification as to why the sponsor should negotiate these costs. This may include the investigator's/site's history of successfully conducting studies, including the ability to recruit assigned goals and perhaps over-enroll, the investigator's/site's reputation for complete and accurate documentation, ability to meet deadlines, preparation for monitoring visits, minimal violations and deviations in previous studies, and other data that indicate the quality and efficiency of the site.

Negotiation, as discussed in the previous chapter, is a two-way street, requiring give and take by both parties. Communication is key—some suggestions include the following:

- Create a list of priorities from which to work.
- Do not start the negotiation with the final acceptable budget—this leaves no room for negotiation.
- As the negotiation proceeds, respond to requests and provide information in a timely manner.

- Ask questions to clarify points not understood.
- Ask questions until site personnel clearly understand the points.

If site personnel are new to negotiation, they may not feel comfortable tackling everything at once. In this case, it is advisable to pick two to three of the site's priority items to negotiate. It is easier to build on successful steps than to recover from a total failure. However, lack of success provides a learning opportunity and allows opportunities for the next negotiation experience. Practice, practice, practice makes for successful negotiation. For example, the site manager might ask that the sponsor reimburse the site for 10 screen failures instead of the sponsor's proposed 5. If the sponsor is not receptive during this negotiation, plan for the next negotiation by collecting additional information from the site's history in the recruitment of similar studies related to the number of screen failures and by providing information on what the site did to prevent or lessen the number of screen failures by implementing prescreening processes.

A VIEW FROM INDIA

As defined by the GCP guidelines published by the Government of India, Ministry of Health & Family Welfare, Central Drugs Standard Control Organization (2004, Section 1), a contract is "A written, dated, and signed document describing the agreement between two or more parties involved in a biomedical study, namely investigator, sponsor, institution. Typically, a contract sets out delegation/distribution of responsibilities, financial arrangements and other pertinent terms. The 'Protocol' may form the basis of 'Contract.'"

As discussed in this chapter, a CTA is a formal and legal document, valid for a defined period of time. The CTA content is similar in India and in the United States as both serve the same purpose. The CTA in India contains

- The details of the parties involved
- A confidentiality clause
- The scope of work and the obligations of each of the signatories
- The expectations from the parties involved
- The monitoring plan
- Data recording and access
- Audits and inspections
- Treatment of injuries/adverse events arising due to participation in the study
- Indeminification
- Jurisdiction
- Term of the contract
- Termination of the contract
- Handling of essential documents and information
- Publication
- The financial plans (payment amount and payment pattern)

A CTA between a sponsor/clinical research organization (CRO) and an investigator is called a bipartite agreement, whereas a CTA among a sponsor/CRO, the investigator, and the research site/institution is a tripartite agreement.

In addition to the points mentioned in this chapter, there are a few others that need mention here from the Indian perspective. Study site personnel should be vigilant about them. They must read the CTA carefully to ensure that the following points are covered:

1. There is a clear mention of number of subjects to be enrolled by the site and the time frame for the enrolling them.
2. The sponsor takes the responsibility for overall study; data handling, verification, and statistical analysis; and preparation of clinical study report.
3. The sponsor will be responsible for providing insurance for the study subjects.
4. The sponsor will be responsible to timely notify the health authorities in India.
5. The details regarding publication rights in case of a multicenter trial are clearly specified.

Understanding the CTA can help avoid issues like frequent amendments, payment delays, and unscheduled or underprepared monitoring visits.

CHAPTER REVIEW

In this chapter, we covered the key terms and clauses of the CTA. The chapter provides a description of the clauses and language that should be included in each clause to protect the investigator/site and the research subject. Also covered were common issues sites experience related to the CTA and a brief discussion of how to negotiate a fair and balanced contract.

APPLY YOUR KNOWLEDGE

1. You have reviewed the sponsor's contract and discovered the following language related to the reporting of protocol violations, "The investigator shall immediately report in writing any violation of the protocol that occurs at the study site." Is this language acceptable to your site? Why or why not? What changes, if any, would you request?
2. Contract termination describes how a contract can be terminated. As a site, what would you need included in this section to protect your site?

REFERENCES

Government of India, Ministry of Health & Family Welfare, Central Drugs Standard Control Organization. (2004). Good clinical practice for clinical research in India. Retrieved from http://www.cdsco.nic.in/html/GCP1.html

Model Agreements and Guidelines International (MAGI). (2016, August 27). MAGI's Clinical Trial Agreement Template v1.1. Retrieved from https://magiworld.org/standards/

Pfeiffer, J. P. (2014). *Clinical trial agreements: Negotiation and management.* Germany: Scholar's Press.

U.S. Department of Health and Human Services, Food and Drug Administration. (2010). *Information sheet guidance for sponsors, clinical investigators, and IRBs: Frequently asked questions—Statement of investigator.* Retrieved from http://www.fda.gov/downloads/RegulatoryInformation/Guidances/UCM214282.pdf

Webb, T. (2006). A checklist for clinical trial agreements. *Journal of Research Best Practices 2*, 1–3.

Windschiegl, M. (2015, November 5–6). Contracting 101 [Presentation slides]. PFS Clinical. Coverage Analysis Workshop, Scottsdale, AZ.

9 U.S. Clinical Trials—Additional Topics

As clinical trials evolve with the innovation of new medical products, various processes and perspectives are also changing. This chapter covers some of the special situations that might be encountered by study sites as they conduct clinical trials in the United States. Topics reviewed in this chapter include adaptive clinical trials, stem cell research, combination products, and biospecimen collection and management.

ADAPTIVE CLINICAL TRIALS

Adaptive trials allow sponsors to modify certain aspects of a clinical trial during the term of the study. The sponsor must review the protocol to determine what aspects of the study are appropriate for adaptation without affecting the integrity of the study. Planning adaptations in advance, as opposed to making protocol amendments during the study, can improve the efficiency of the study and reduce the time it takes to complete the study. Carefully planned adaptive designs are appropriate for phase 1, 2 and 3 trials. According to the U.S. Food and Drug Administration (FDA, 2010, p. 2) "... an *adaptive design clinical study* is defined as a study that includes a prospectively planned opportunity for modification of one or more specified aspects of the study design and hypotheses based on analysis of data (usually interim data) from subjects in the study. Analyses of the accumulating study data are performed at prospectively planned time points within the study, can be performed in a fully blinded or in an unblinded manner, and can occur with or without formal statistical hypothesis testing." Changes in design or analysis through the examination of cumulated data at interim points throughout the study can lead to more efficient studies, which may require fewer patients, reduce length of the study, and increase the chance of answering the study question correctly (Kairalla, Coffey, Thomann, & Muller, 2012; Shein & Chang, 2008, U.S. FDA, 2010).

Carefully planned adaptive designs allow the use of cumulative data to modify the design after the trial initiation without affecting the validity and integrity of the trial (Shein & Chang, 2008). Possible modifications must be determined prospectively and included in the protocol or statistical analysis plan. Examples of modifications that might be planned in the protocol include the following:

- Study eligibility criteria (either for subsequent study enrollment or for a subset selection of an analytic population)
- Randomization procedures
- Treatment regimens of the different study groups (for example, dose level, schedule, duration)
- Total sample size of the study (including early termination)

- Concomitant treatments used
- Planned schedule of patient evaluations for data collection (for example, number of intermediate time points, timing of last patient observation, and duration of patient study participation)
- Primary endpoint (for example, which of several types of outcome assessments, which time point of assessment, use of a unitary versus composite endpoint, or the components included in a composite endpoint)
- Selection and/or order of secondary endpoints
- Analytic methods to evaluate the endpoints (for example, covariates of final analysis, statistical 115 methodology, Type I error control) (U.S. FDA, 2010, p. 7)

As clinical trials continue to become more complex and sponsors look to reduce or contain costs by creating more efficient trials using fewer subjects and decreasing time to market for new medical products, the use of adaptive designs may be adopted more frequently in clinical trials.

COMBINATION PRODUCTS

Combination products are medical devices that include a device and a drug or biologic, each of which is regulated separately, but for the intended use, both are required in combination. These can be combined into one product such as a prefilled syringe or packaged separately and combined for the intended use, for example, a surgical kit.

Under 21 C.F.R. 3.2(e), a combination product includes the following:

- A product composed of two or more regulated components, that is, drug/device, biologic/device, drug/biologic, or drug/device/biologic, that are physically, chemically, or otherwise combined or mixed and produced as a single entity (for example, a prefilled syringe or drug-eluting stent);
- Two or more separate products packaged together in a single package or as a unit and composed of drug and device products, device and biological products, or biological and drug products (for example, a surgical or first-aid kit);
- A drug, device, or biological product packaged separately that according to its investigational plan or proposed labeling is intended for use only with an approved, individually specified drug, device, or biological product where both are required to achieve the intended use, indication, or effect and where upon approval of the proposed product the labeling of the approved product would need to be changed (for example, to reflect a change in intended use, dosage form, strength, route of administration, or significant change in dose) (a "cross-labeled" combination product, as might be the case for a light-emitting device and a light-activated drug); or
- Any investigational drug, device, or biological product packaged separately that according to its proposed labeling is for use only with another

individually specified investigational drug, device, or biological product where both are required to achieve the intended use, indication, or effect (another type of cross-labeled combination product) (FDA Product Jurisdiction, 2015).

The FDA established the Office of Combination Products (OCP) in December 2002 as required by section 204 of the Medical Device User Fee and Modernization Act of 2002. The primary responsibilities of the OCP are to ensure the prompt assignment of combination products to agency centers and to oversee their "timely and effective" premarket review and "consistent and appropriate" postmarket regulation. The OCP develops policy, guidance, and regulations for combination products and works with the other centers in the approval process. The OCP assists in determining which center has jurisdiction for approval of an investigational combination product and provides oversight of the premarket review and ensures consistency in the review and postmarket regulations by all centers. It also serves as a resource for the industry.

As mentioned earlier, combination products are composed of components that, individually, would usually be regulated by separate centers of the FDA under different types of regulatory authorities (i.e., biologics, drugs, and devices). Each of the three review centers are involved in the assessment of combination products. The Center for Biologics Evaluation and Research, the Center for Drug Evaluation and Research (CDER), and the Center for Devices and Radiological Health (CDRH) each maintains a comprehensive, regularly updated section of the FDA website that provides detailed information describing their specific organizational structures, how to contact each of the different centers, and advice as to how applications should be prepared for submission to the respective centers. Dealing with three centers, each having its own regulations and authorities, can prove to be quite challenging when seeking approval of a combination product.

A combination product is assigned to the center that has primary jurisdiction for premarket review. Primary jurisdiction is based on the determination of which constituent part of the investigational medical product provides the primary mode of action (PMOA). The OCP determines PMOA and the center of jurisdiction. The PMOA is the therapeutic action expected to make the greatest contribution to the overall effect of the combination product (Elser, 2016). For example, the CDRH will be the lead agency for bone-cement-containing antibiotics, as the PMOA is the cement that provides stabilization of the bone. An asthma inhaler falls under the jurisdiction of CDER, as the PMOA is the medication to open the bronchial airways.

Drugs are regulated under 21 C.F.R. 210 and 211; biologics, under 600 and the Public Health Services Act Part F, subpart 1; and devices, under 800. 21 C.F.R. Part 4 brings these regulations together to manage combination products. Part 4 sets forth a transparent and streamlined regulatory framework when demonstrating compliance for "single entity" and "copackaged" combination products (Elser, 2016; U.S. FDA, 2016, 21 C.F.R. Part 4). Agency centers will coordinate as appropriate regardless of who is the lead to ensure compliance with current good manufacturing practices (CGMP) regulations outlined in 21 C.F.R. Part 4.

Regulations require that studies for combination products include evaluating not only the impact of the device on the drug/biologic but also the impact of the product on the device. In addition, human factor studies must be conducted to demonstrate that the combination product works for the intended patient population. The population must be able to use the product and the product must work. (For example, a prefilled syringe containing medication for rheumatoid arthritis [RA] may not be appropriate for that patient population due to the effects of the disease on the joints of the fingers. The patient may not be able to hold the syringe or to push the plunger.)

Examples of combination products include the following:

- Prefilled syringes
- Drug eluting coronary stents
- Drug and transdermal patches
- Surgical mesh with antibiotic coating
- Inhalers
- Insulin injection pumps
- Alcohol swabs
- Drug coated devices such as catheters
- Bone cement containing antibiotics
- Kits—epidural tray, first aid kits
- Drug or biologic packaged with a delivery device

Manufacturers of combination products must demonstrate compliance with applicable CGMP requirements and quality system regulations for both components of the combination product. Methods, facilities, and manufacturing controls also must be applied to the development of combination products. The manufacturer must understand the regulations that apply to each component of the combination product and the differences between drugs and devices to ensure compliance of all required regulations during the development process.

The premarket submission of combination products provides many challenges for the manufacturer. According to John Barlow Weiner, Associate Director of Policy at OCP (2013), as quoted in "Regulatory Compliance," combination products face many unique challenges. These challenges include the following:

- Legal—there are currently no statutory or regulatory standards for combination products.
- Marketing—the drug, biological product, and device industries differ in their markets.
- Premarket—drugs and devices have different data requirements, review timelines, and other premarket factors such as promotion and advertising (i.e., what can be said or claimed about a device versus a drug).
- Postmarket—challenges included clarity of duties and reconciling regulatory requirements.
- Cross-labeling—these challenges, which can be unique to combination products, include labeling consistency, proprietary data reliance, coordination of changes, coordination of marketing review, and authorization (Wiltz, 2013).

In addition, in 2015, the FDA's Office of Planning conducted a Combination Product Review Intercenter Consult Process Study to examine the current procedures and processes for premarket submissions of combination products. The study found the following:

- Differing review approaches and review timelines of the three centers inhibit effective coordination between the centers.
- Lack of a shared technical platform inhibits the ability to access information in a timely manner by the different centers resulting in communication gaps.
- There is unclear communication between the centers.
- Resources are not adequate. Centers are not able to reimburse the consulting time of staff from other centers stretching already limited resources (Department of Health and Human Services, 2015).

These issues and challenges and recommendations from the study will be reviewed by the agency. It is expected that revisions to the process will be proposed and guidance documents updated in upcoming years.

STEM CELL RESEARCH

Stem cells are the foundation cells for human bodies, making up the body's organs and tissues. Stem cells originate from an initial pool of stem cells that are formed shortly after an egg is fertilized during reproduction. Stem cells are defined by two characteristics: their capacity to (1) self-renew (divide in a way that generates more stem cells) and (2) to differentiate (to turn into mature, specialized cells that make up our tissues and organs) (International Society for Stem Cell Research, n.d.).

Pluripotent stem cells or human embryonic stem cells (hESCs) are derived from human embryos or human fetal tissue. Induced pluripotent stem cells (iPSCs) are adult cells that have been genetically reprogrammed to an embryonic stem-cell-like state by being forced to express genes and factors important for maintaining the defining properties of embryonic stem cells (ESCs; National Institutes of Health [NIH], 2015).

Studying stem cells provides researchers with a better understanding of the human body. Researchers hope to find the answers to the diagnosis, treatment, or prevention of many diseases and conditions such as heart disease, stroke, spinal cord injury, Parkinson's disease, Huntington's disease, hearing loss, RA, and many types of cancer, through stem cell research. Stem cells can be used to search for new drugs and to improve drug functions and may help in the development of targeted medicine, prevent medication complications, and pave the way to regenerative medicine (U.S. NIH, 2015).

In the United States, stem cells are regulated under 21 C.F.R. 1271 and the Public Health Services Act (PHS Act) Section 361. This section of the code regulates the registration of establishments that manufacture human cells, tissues, and cellular and tissue-based products (HCT/P's). It also establishes donor eligibility, current good tissue practices (cGTPs), and other procedures to prevent the introduction, transmission, and spread of communicable diseases. Section 361 of the PHS Act (2003) authorizes the FDA to "make and enforce such regulations as...are necessary to prevent the introduction, transmission, or spread of communicable diseases." Section 361 of the PHS Act also refers to 21 C.F.R. 45 and 46 and the protection and safeguard of human research subjects who participate in clinical research, including informed consent and institutional review board (IRB) review (PHS Act, Section 361).

In March 2009, President Obama signed Executive Order 13505 "Removing Barriers to Responsible Scientific Research Involving Human Stem Cells," directing the NIH to issue a guidance document on human stem cell research within 120 days (Exec. Order No. 13505, 2009). On July 6, 2009, the NIH issued the final "Guidelines for Stem Cell Research." These guidelines support the principle that ESC research offers the possibility to better understand the disease process and to provide better ways to diagnose and treat diseases. These guidelines also address the need for proper voluntary informed consent of individuals providing embryos for research.

The NIH Guidelines permit federal funding of research for NIH registered hESCs as well as identify categories of stem cell research that are ineligible for NIH funding. Categories that do not qualify for funding include the following:

- Stem cells that are "introduced into non-human primate blastocysts"

or

- Research that involves the "breeding of animals where the introduction of hESCs...or human induced pluripotent stem cells may contribute to the germ line" (U.S. NIH, 2009).

The NIH guidelines state that federal funding will not be available for research involving stem cells that are derived from sources such as therapeutic cloning. The guidelines also make reference to the Dickey-Wicker Amendment, which prohibits the funding of stem cells from human embryos and bans the use of federal funds for the creation of human embryos for research purposes. In addition, the amendment bans research in which human embryos are "destroyed, discarded, or knowingly subjected to risk of injury or death greater than that allowed" for federally funded research on fetuses in utero (U.S. NIH, 2009).

Stem cell research is new and rapidly changing and regulations are still catching up. Clinical trials using stem cells must comply with all applicable current regulations that govern human subject research. Protocols and study documents must be reviewed and approved by an IRB. As with all research, potential subjects should be thoroughly informed during the consent process to ensure they understand the intricacies and implications of the research. This may take more time and patience on the part of the investigator, especially if the concepts, ethical implications, and long-term impact of these studies are novel to subjects. If the research involves use of the subject's stem cells, a separate consent is usually required for future use of these stem cells for other research.

As stem cell research grows, regulations and guidance documents will be developed and incorporated into the human subject research regulations. It is important that research sites that are involved in stem cell research or may be in the future remain current with changes to regulations and guidance documents related to this research.

BIOSPECIMEN COLLECTION AND REPOSITORIES

Many current clinical trials include, as part of the study or as a separate substudy, the collection of biological specimens that are used in analyzing disease processes, investigational product effectiveness, and other current and future research goals. These samples are termed *biospecimens* and are defined as "samples of material, such as urine, blood, tissue, cells, DNA, RNA, and protein from humans, animals, or plants" (U.S. NIH, National Cancer Institute, 2014).

Biospecimens that are to be used for future research are stored in a biorepository. Biorepositories are equipped with special equipment to store the specimens such as −40°C freezers. Biorepository software is utilized to inventory, track the location, and manage the specimen and specimen information for future use. Researchers may analyze these samples for indications of disease, to provide a better understanding of disease progression, biological pathways, translational research, and precision medicine. Collection, processing, analyzing, and using the biospecimens for research are a complex process that involves multiple stakeholders. These stakeholders include the individual consenting and collecting the sample, the local lab processing the sample for shipment to the repository, and the various biobank personnel.

Generally, participation in a biospecimen substudy is voluntary and in no way affects the subject's participation in the main study. Exceptions to this might include oncology studies. Collecting biospecimen samples for future research requires permission from the donor. This is done through informed consent. As discussed in Chapter 4, the informed consent is more than a form, it is a process.

The clinical trial site is often the first in the line of the stakeholders. Consent must be given prior to collecting any sample. The site must follow the protocol for the collection of the sample to ensure that it is carefully collected and the sample is adequate and meets the requirements as specified in the protocol. The sample must be processed in the site lab according to the protocol and specific lab manual instructions. Packaging and shipping must conform to applicable standards and regulations. These standards are set by the U.S. Department of Transportation and the International Air Transport Association. All personnel involved in the handling and shipping of biospecimen samples should be properly trained.

The consenting process is governed by the Common Rule. These requirements are found in 45 C.F.R. part 46. The Common Rule outlines the basic requirements for IRB review and approval of clinical investigations involving the collection of biological specimens as well as the required components of the informed consent.

At the time of the writing of this book, the FDA has proposed revisions to the Common Rule, including changes to the biospecimen consent, the management of biospecimen collection, processing, and storage.

A VIEW FROM INDIA

ADAPTIVE CLINICAL TRIALS

As mentioned previously in this chapter, an adaptive clinical trial is a trial for which the sponsor prospectively plans modifications for the trial that do not affect the integrity and statistical validity of the trial. With the many stated benefits of adaptive clinical trials, and with a robust regulatory framework, affordable drugs can be developed through adaptive clinical trials not only in India but also globally.

As in the case of all clinical trials in India, the office of the Drug Controller General of India (DCGI) and the respective IRB/independent ethics committee (IEC) are the approving bodies for the adaptive designed clinical trial. The DCGI will approve an adaptive clinical trial if the same trial is already approved by the FDA. However, if not, and because there are no formal guidelines available for approval of adaptive clinical trials in India, the DCGI reviews them on an individual case-by-case basis.

Combination Products

In India, the regulation of combination products is not clearly defined. One definition of a drug found in the Drugs and Cosmetics Act is "such devices intended for internal or external use in the diagnosis, treatment, mitigation or prevention of disease or disorder in human beings or animals, as may be specified from time to time by the Central Government by notification in the Official Gazette, after consultation with the Board" (Government of India, Ministry of Health & Family Welfare [MOHFW], 2005, p. 2).

Some products that are found in the combination product category are treated as drugs irrespective of the device component associated with it. As an example, the MOHFW under Gazette notification S.O. 1468(E), dated 10 June 2005, declared the following sterile devices to be considered drugs under section 3 (b)(iv) of the Act, which are regulated as notified medical devices:

1. Cardiac stents
2. Drug eluting stents
3. Catheters
4. Intra ocular lenses
5. IV cannulae
6. Bone cements
7. Heart valves
8. Scalp vein set
9. Orthopedic implants
10. Internal prosthetic replacements (Government of India, MOHFW, 2010, p. 1)

In 2009, Indian regulators introduced guidelines establishing that the PMOA would be the basis for the regulatory pathway of combination products. However, these regulations do not apply to the products already classified as drugs (notified medical devices). Even though the 2009 regulation classifies new combination products on the basis of PMOA, the Central Drug Standard Control Organization (CDSCO) can still classify any new combination product in the notified medical devices category and hence regulate it as drug.

Stem Cell Research

The Indian Council of Medical Research (ICMR) and the Department of Biotechnology (DBT) established Guidelines for Stem Cell Research and Therapy in 2007. Considering the advancement in the field and the views of the stakeholders, ICMR, the Department of Health Research (DHR), and the DBT revised the 2007 guidelines in December 2013 and renamed them the National Guidelines for Stem Cell Research.

The National Apex Committee for Stem Cell Research and Therapy (NAC-SCRT) is the body constituted by the MOHFW, the DHR, and the Government of India to examine the scientific, technical, ethical, legal, and social issues involving stem cell research and therapy in India. All institutions conducting research on human stem cells must constitute an Institutional Committee for Stem Cell Research (IC-SCR) and mandatorily register it with the NAC-SCRT. The NAC-SCRT and IC-SCR will oversee and monitor activities in the field of stem cell research at the national and institutional levels, respectively.

The stem cell research guidelines established three categories of stem cells based on the cell type or tissue of origin: somatic stem cells, ESCs, and iPSCs.

As stated in the National Guidelines for Stem Cell Research–2013, hematopoietic stem cell transplantation (HSCT) for hematological disorders is the only approved indication for stem cell therapy. All stem cell therapies other than HSCT are treated as investigational and must be conducted only in the form of clinical trials. As per the guidelines, all clinical trials involving stem cells must be conducted in compliance with Schedule Y of the Drug & Cosmetics Act 1940 & Drug and Cosmetic Rules 1945 (India), CDSCO's Good Clinical Practice guidelines, and ICMR's Ethical Guidelines for Biomedical Research Involving Human Participants. The guidelines state the responsibilities of the investigator, institution, and the sponsor in conducting stem cell research. The protocol format is given in Annexure II of the guidelines. Clinical trials using stem cells requires an approval from DCGI after obtaining approval from NAC-SCRT through IC-SCR and IRB/IEC.

All clinical trials involving the use of stem cells must be registered with the Clinical Trial Registry of India. For monitoring of clinical trials in addition to the regular committees, a separate Data Safety and Monitoring Board should be constituted for each trial. All records pertaining to clinical adult stem cell research and ESCs/iPSCs research must be maintained for a period of 5 and 10 years, respectively.

As quoted from a press release, on February 21, 2014, the MOHFW vide order dated September 1, 2010, constituted a Core Investigational New Drug Panel of Experts, namely, Cellular Biology Based Therapeutic Drug Evaluation Committee (CBBTDEC) under the chairmanship of Director General, ICMR & Secretary, DHR, to advice the DCGI in matters pertaining to regulatory pathways leading to the approval of clinical trials and market authorization for the "Therapeutic products derived from Stem Cell, human Gene manipulation and Xenotransplantation technology." The CBBTDEC has deliberated on the need for strengthening the regulatory agency (DCGI) by establishing a separate wing for stem cell research supported with knowledge and capacity to regulate the activities in the country. Accordingly, the Directorate General of Health Services, the office of Drugs Controller of General

(India) MOHFW, and the Government of India vide order dated March 16, 2012, has established a Stem Cell Division within the Biological Division in the CDSCO for the internal evaluation of all proposals including stem cell concerning clinical trials and marketing authorization before referring to CBBTDEC (Government of India, Ministry of Health & Family, 2014).

CHAPTER REVIEW

This chapter covered some of the special situations that a site or researcher may encounter while doing human subject research. The chapter provided a brief overview of the topics to provide the reader with a general understanding of these situations. As mentioned, the regulations and guidelines for biospecimens are currently being revised and stem cell research guidelines continue to be developed.

APPLY YOUR KNOWLEDGE

1. Executive Order 13505 removed the ban from using stem cells for research and directed the NIH to provide guidelines for the use of stem cells in research. The guidelines developed by NIH permit the use of hESC and identify categories of stem cells that will not be eligible for federal funding. What are the two categories of stem cells that are not eligible for federal funding? What do you think was NIH's reasoning for not including these two categories?
2. As discussed in this chapter, a lead center will be assigned by the OCP to review and approve combination products based on the product's PMOA. Which center has jurisdiction for each of the following medical products and why?
    a. Insulin injection pump
    b. Drug eluting coronary stent
    c. Transdermal patch

REFERENCES

Department of Health and Human Services. (2015). *Combination product review: Intercenter consult process study*. Retrieved from http://www.fda.gov/downloads/CombinationProducts/GuidanceRegulatoryInformation/UCM467128.pdf

Elser, C. (2016). *Combination products: An overview* [Presentation slides]. Retrieved from Arizona State University, College of Nursing and Health Innovation: https://nursingandhealth.asu.edu/crm-webinar-series

Exec. Order No. 13505, 74 Fed. Reg. 10667 (2009).

FDA Product Jurisdiction, 21 C.F.R. § 3.2(e) (2015).

Government of India, Ministry of Health & Family Welfare. (2005). *The Drug and Cosmetics Act and Rules*. Retrieved from http://www.cdsco.nic.in/writereaddata/Drugs&CosmeticAct.pdf

Government of India, Ministry of Health & Family Welfare. (2010). *List of notified medical devices*. Retrieved from http://cdsco.nic.in/Medical_div/list%20of%20notified%20medical%20device.0001.pdf

Government of India, Ministry of Health & Family Welfare. (2013). *The drugs and Cosmetics Act, 1940 and Rules, 1945 Schedule Y.* Retrieved from http://www.mohfw.nic.in/WriteReadData/l892s/43503435431421382269.pdf

Government of India, Ministry of Health & Family. (2014). National guidelines for stem cell research. Retrieved from http://pib.nic.in/newsite/PrintRelease.aspx?relid=104095

International Society for Stem Cell Research. Stem cell facts. Retrieved from http://www.isscv.org/visitor-types/public/stem-cell-facts

Kairalla, J. A., Coffey, C. S., Thomann, M. A., & Muller, K. E. (2012). Adaptive trial designs: A review of barriers and Opportunities. Trials 2012, 13:145. Retrieved from http://www.trialsjournal.com/content/13/1/145

Shein, C. C., & Chang M. (2008). Adaptive design methods in clinical trials—A review. *Orphanet Journal of Rare Diseases*, *3*, 11. Retrieved from http://www.ojrd.com/content/3/1/11

U.S. Food and Drug Administration. (2010). *Guidance for industry: Adaptive design clinical trials for drugs and biologics.* Retrieved from http://www.fda.gov/downloads/Drugs/GuidanceComplianceRegulatoryInformation/Guidances/UCM201790.pdf

U.S. Food and Drug Administration. (2016). 21 CFR Part 4. Regulation of Combination Products. Retrieved from http://www.accessdata.fda.gov/scripts/cdrh/CFRSearch.cfm?CFMPart=4

U.S. National Institutes of Health. (2009). National Institutes of Health guidelines on human stem cell research. Retrieved from http://stemcells.nih.gov/policy/2009-guidelines.htm

U.S. National Institutes of Health. (2015). Stem cell basics. Retrieved from http://stemcells.nih.gov/info/basics/Pages/Default.aspx

U.S. National Institutes of Health, National Cancer Institute. (2014). Patient corner: What are biospecimens and biorepositories? Retrieved from http://biospecimens.cancer.gov/patientcorner/

Wiltz, C. (2013, April 11). A peek at challenges from the Office of Combination products [blog post]. DeviceTalk. Retrieved from http://www.mddionline.com/blog/devicetalk/peek-challenges-office-combination-products

10 Clinical Research and India

India has always fascinated the United States and Europe. In its earlier years, it was a country of snake charmers, but more recently, India's software programmers have forced the world to take notice. The clinical research industry is a part of this technology-oriented vision for India. Compared with the United States, the clinical research industry is small but is rapidly growing. India's patent laws are closely aligned with U.S. laws and the clinical research procedures in India are being standardized to global norms. As a result, most major global clinical research organizations have an office in India, and India is fast becoming a favored destination for many international pharmaceutical firms.

In this chapter, we discuss the clinical research industry and practices in India. The chapter begins with a brief introduction to the pharmaceutical industry and then we discuss India as a clinical research destination for global firms. In the next section, we look at the regulatory framework. Finally, we present a current environment of research in India.

PHARMACEUTICAL/MEDICAL INDUSTRY IN INDIA

India's pharmaceutical industry was ranked third globally in terms of volume (Panchal, Kapoor, & Mahajan, 2014), with an estimated value of approximately $18 billion in December 2015 (PTI, 2015). In 2014, pharmaceutical exports from India were around $11.7 billion, with a third of these exported to the United States (India Exporter: India, 2014). The number of pharmaceutical firms from India registered with the Food and Drug Administration (FDA) is higher than from any other country. Additionally, the industry is slated to grow annually at close to 8% for the next 5 years (PTI, 2015).

A key turning point in the Indian pharmaceutical industry was the implementation of the product patent regime in January 2005, making India completely compliant with Trade-Related Aspects of Intellectual Property Rights. Previously, Indian firms had thrived on reverse engineering drugs patented in the United States and other countries. The new law forced the pharmaceutical industry to change its practices and find new ways to remain competitive. As a result, the stability of the patent laws and a free and fair judiciary has motivated global firms to do business in India.

India has a robust and high-quality medical school system, with 50,000 doctors joining the medical fraternity every year. Currently, there are about 1 million registered doctors. India is the largest "exporter" of medical practitioners to the United States. Because of its experienced medical professionals, the existence of world class infrastructure, and lower costs, a booming medical tourism industry has emerged. In 2015, the medical tourism industry was valued at $3 billion. The Medical Tourism

Market Report: 2015 mentioned that India offered a variety of medical procedures at one-tenth the cost in the United States at equal quality levels.

CLINICAL RESEARCH INDUSTRY IN INDIA

India has a population of more than 1.2 billion people and has 20% of the global disease burden (National Commission on Macroeconomics and Health, 2005). The disease burden is composed of both noncommunicable and communicable diseases.

As of June 2015, India accounted for only 1.4% (2,650 studies out of 191,938) of all global clinical trials (Kafaltia & Manchanda, 2015, p. 39). Most of the studies in India are phase 2 and 3 studies. India conducted 7% of all global phase III and 3.2% of all global phase II trials, according to a 2012 study by Ernst & Young and the Federation of Indian Chambers of Commerce and Industry (2009).

INDIA: A FAVORED GLOBAL DESTINATION FOR CLINICAL RESEARCH

India is appealing to clinical research sponsors throughout the world for many reasons, including its diverse population, its regulatory environment, its reasonable cost of living, and its English-speaking population. In the following paragraphs, we look at some of these aspects:

Population

India has a large population, which is also genetically diverse. All major diseases of the developed and the developing worlds are prevalent. India has a highly educated and extremely efficient medical fraternity and trained work staff. To augment the trained staff, super-specialty hospitals with state-of-the-art facilities exist in both the public and private sectors. India also has robust experience in conducting clinical trials per global standards due to the high number of clinical trials conducted in the country. Information technology systems for data management and documentation are well developed and data scientists and statistical support are easily accessible. Additionally, electronic case report forms are widely used throughout clinical sites.

Regulatory Environment

There is a well-defined and robust regulatory system in place. The regulatory authorities work closely with international bodies to align Indian regulations with the global standards. With the support and experience of international regulatory agencies, the Indian regulatory system has now become more efficient.

Cost of Living

Due to the lower cost of living in India, the overall cost for conducting high-quality clinical trials is significantly lower than the cost in the United States and the European

Union (EU). In India, the clinical trial cost is as low as 0.11 units compared with 1.0 unit in the United States (Mondal & Abrol, 2015). For example, a hepatitis drug in the United States costs $1,000, whereas in India, a generic version of the same drug is available for $4.00 (Gokhale, 2015).

The low cost does not imply a deficiency in the quality of clinical trials. As mentioned, the doctors are highly qualified, the hospitals are world class, and regulations are being updated to reflect global standards. The use of English as a primary language adds to the already strong position. These factors continue to make India a desirable global destination for conducting clinical trials.

REGULATORY FRAMEWORK

The Central Drugs Standard Control Organization (CDSCO) in India is equivalent to the FDA of the United States and the European Medicines Agency of the EU. The CDSCO falls under the Directorate General of Health Services (Dte GHS) and the Ministry of Health & Family Welfare (MOHFW), Government of India.

The Drugs Controller General of India (DCGI) is the head of the CDSCO. The DCGI is the regulatory authority for approving new drugs and medical devices and clinical trials, as well as monitoring the quality, efficacy, and safety of pharmaceutical products on the market.

The DCGI is advised by the Drugs Technical Advisory Board (DTAB) and the Drug Consultative Committee (DCC). The DTAB advises on technical matters arising out of the administration of the Drugs and Cosmetics Act, 1940, and Rules, 1945 (D & C Act), whereas the DCC advises on the matters for uniformity on the administration of the D & C Act (Figure 10.1).

National Apex Committee for Stem Cell Research and Therapy

The MOHFW, along with the Department of Health (DHR) and government of India, formed the National Apex Committee for Stem Cell Research and Therapy to oversee scientific, technical, ethical, legal, and social issues involving stem cell research and therapy in India.

Indian Council of Medical Research

The Indian Council of Medical Research (ICMR) is the apex body for the planning, formulation, coordination, implementation, conduct, and promotion of biomedical research in India. It is one of the oldest medical research bodies in the world.

Department of Biotechnology

The Department of Biotechnology (DBT) falls under the Ministry of Science and Technology and is responsible for development and commercialization in the field of modern biology and biotechnology in India. It was set up in 1986.

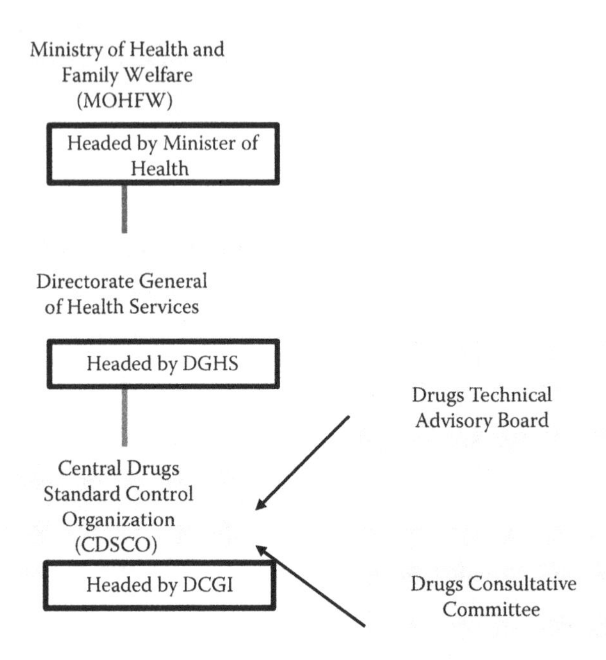

FIGURE 10.1 Organizational chart regulatory authorities—India.

RULES, REGULATIONS, AND GUIDELINES

Regulations are systems to ensure that the quality and integrity of data collected in clinical trials are maintained and the rights, welfare, and safety of research participants are protected.

The D & C Act, amended in 2013, defines a clinical trial as "a systematic study of new drug(s) in human subject(s) to generate data for discovering and/or verifying the clinical pharmacological (including pharmacodynamic and pharmacokinetic) and/or adverse effects with the objective of determining safety and/or efficacy of the new drug" (MOHFW, 2013, rule 122-DAA).

India's regulatory oversight can briefly be divided into law, regulations, and guidelines.

Laws: D & C Act

Clinical trials in India are regulated in accordance with the rules and regulations contained in Part X-A (Import or Manufacture of new drug for clinical trials or marketing) of the D & C Act. The D & C Act has been amended from time to time since its inception. Schedule Y for clinical research was added to the D & C Act in 1988. It contains powers for regulating and ensuring quality, safety, and efficacy of drugs and clinical trials. Clinical trials in India are conducted under Rule-122DA, 122DAA, 122DAB, 122DAC, 122DD, and 122E and Schedule Y of the D & C Act.

Regulations: Schedule Y

Schedule Y is equivalent to the IND regulations described in 21 C.F.R. Part 312. Schedule Y covers the requirements and guidelines for permission to undertake clinical trials under part X-A of the D & C Act. Schedule Y outlines extensive study criteria in line with the globally accepted formats such as International Conference of Harmonisation (ICH) Guidelines and U.S. FDA Regulations. Schedule Y has three major sections and 12 appendices in the 2013 amended version of the D & C Act. An application for approval to conduct a clinical trial is submitted to the DCGI as per Form 44 of the D & C Act and accompanied by data as detailed in Appendix I of Schedule Y. Schedule Y stipulates the responsibilities of the sponsor, investigator, and the ethics committee (EC). Structure, content, and formats for protocols, study reports, EC approvals, informed consent form, and serious adverse event reporting are incorporated.

Guidelines

Guidelines prevalent in the Indian system include the following:

- The ICMR issued the Ethical Guidelines for Biomedical Research on Human Subjects in 2000, which was revised in 2006. The guidelines refer to the Council for International Organizations of Medical Sciences and ICH-Good Clinical Practice (GCP). Most recently, the ICMR has published the draft of the "National Ethical Guidelines for Biomedical and Health Research on Human Participants 2016," which covers new areas, such as Social and Behavioral Sciences, Responsible Conduct of Research, and New Technologies. Additionally, other specialized areas, including the informed consent process, biological materials and datasets, vulnerability, international collaboration, research during humanitarian emergencies, and disasters, have been expanded.
- CDSCO released Indian GCP guidelines in 2001. These guidelines have evolved with consideration of the World Health Organization, ICH, U.S. FDA, and European GCP guidelines as well as the Ethical Guidelines for Biomedical research on Human Subjects issued by ICMR. The DTAB has endorsed adoption of this guideline for streamlining the clinical studies in India.
- ICMR, DHR, and DBT established Guidelines for Stem Cell Research and Therapy in 2007, which were revised in 2013 as the National Guidelines for Stem Cell Research (NGSCR). The aim and scope of the NGSCR are to lay down general principles keeping in view the ethical issues for stem cell research and therapy and to formulate specific guidelines for derivation, propagation, differentiation, characterization, banking, and use of human stem cells.

The Clinical Trials Registry–India (CTRI), hosted at the ICMR's National Institute of Medical Statistics and launched in July 2007, is a free and online public record system for registration of clinical trials being conducted in India. Initiated as a voluntary measure, it became mandatory in June 2009. Moreover, editors of biomedical journals of 11 major journals of India have declared that only registered trials would be considered for publication. The sponsor or the principal investigator can register the clinical trial on CTRI.

All clinical trials involving drugs, medical devices, and stem cells must be conducted in compliance with Schedule Y of the D & C Act, CDSCO's GCP guidelines, ICMR's Ethical Guidelines for Biomedical Research involving Human Participants, and NGSCR and must be registered with CTRI.

CURRENT ENVIRONMENT

The Indian regulatory authorities have taken many aggressive steps to ensure quality and integrity in the clinical research area. They continue to upgrade current policies and to institute new regulations with more stringent measures to monitor compliance in the conduct of clinical research activities.

Some of the noteworthy changes made by the regulatory authorities of India in recent times include but are not limited to the following:

- Registration of all clinical trials on CTRI.
- Registration of institutional review boards (IRBs)/independent ethics committee with the licensing authority (DCGI; as per gazette of India notification G.S.R. 72 (E) dated February 8, 2013, under Rule 122DD).
- Recommendation for conduct of clinical trials with accredited investigators, institutes, and clinical study sites. Organizations like the Quality Council of India and its National Accreditation Board for Hospitals (NABH) and health care providers have designed extensive healthcare standards for hospitals and healthcare providers. The NABH is equivalent to the National Committee for Quality Assurance in the United States.
- Permission to conduct clinical trials (as per gazette of India notification G.S.R. 63 (E) dated February 1, 2013, under Rule 122DAC).
- Audio-video documentation of consent in specific trials with special populations (as per gazette of India notification G.S.R. 611 (E) dated July 31, 2015).
- Compensation in case of injury or death of the subject due to participation in a clinical study (as per gazette of India notification G.S.R. 53 (E) dated January 30, 2013, under Rule 122DAB).
- Inspection of clinical trial sites (as per gazette of India notification G.S.R. 572 (E) dated July 17, 2012 under Rule 122DAC (h)).
- Applications to the DCGI include the following information, per the September 2014 DCGI order:
  - "Assessment of risk versus benefit to the patients
  - Innovation vis-a-vis to existing therapeutic option
  - Unmet medical need in the country (India)" (Dte GHS, 2014, p. 1).

For new drug substances discovered and developed in countries other than India, phase I data, which are generated outside India, are required to be submitted along with the application to conduct a clinical trial (Form 44 of D & C Act as mentioned previously). Permission to carry out clinical trials in India is generally given in stages, taking into consideration the data emerging from earlier phase(s). Based on the application and the data submitted, permission to repeat phase I trials and/or conduct phase II trials and subsequently phase III trials concurrently with other global trials of that drug may be granted. If the drug is intended for marketing in India, phase III trials must be conducted in India to receive permission to market.

CONCLUSION

Clinical investigators, sponsors, and regulatory bodies play a critical role in ensuring high-quality clinical studies. The government of India and the regulatory authorities are developing and implementing a robust regulatory system to ensure that the clinical trials conducted in India are of high quality and integrity, allowing high-quality and reasonably priced medicines to reach its masses. The regulatory authorities have amended existing regulations and have put new guidelines in place to strengthen the clinical research industry.

Slowly but surely, the regulators will achieve their goal of enhancing the clinical research industry—and India will continue to become a global leader in clinical research.

REFERENCES

Directorate General of Health Services. (2014). *Order 132*. Retrieved from http://www.cdsco.nic.in/writereaddata/Office%20Order%20dated%2005_09_2014.pdf

Ernst & Young/FICCI. (2009). *The glorious metamorphosis; Compelling reasons for doing clinical research in India*. Retrieved from http://www.ficci.com/publication-page.asp?spid=20026

Gokhale, K. (2015). The same pill that costs $1,000 in America sells for $4 in India. Retrieved from http://www.bloomberg.com/news/articles/2015-12-29/the-price-keeps-falling-for-a-superstar-gilead-drug-in-india

Government of India, Ministry of Health & Family Welfare. (2013). *The Drugs and Cosmetics Act, 1940 and Rules, 1945 Schedule Y*. Retrieved from http://www.mohfw.nic.in/WriteReadData/l892s/43503435431421382269.pdf

India Exporter: India. (2014). *World's Richest Countries*. Retrieved from http://www.worldsrichestcountries.com/top_indian_exporters_trade_partners.html

Kafaltia, M., & Manchanda, M. (2015). Clinical research in India—Current scenario and the need of the hour. *Applied Clinical Research, Clinical Trials & Regulatory Affairs, 2*, 37–45. http://dx.doi.org/10.2174/2213476X02666150818221707

Mondal, S., & Abrol, D. (2015). *Clinical trials industry in India: A systematic review*. Institute for Studies in Industrial Development 179. Retrieved from Institute for Studies in Industrial Development: http://isid.org.in/pdf/WP179.pdf

National Commission on Macroeconomics and Health. (2005). *Burden of disease in India*. Retrieved from http://www.who.int/macrohealth/action/NCMH_Burden%20of%20disease_(29%20Sep%202005).pdf

Panchal, M., Kapoor, C., & Mahajan, M. (2014). Success strategies for Indian pharma industry in an uncertain world. Retrieved from http://www.business-standard.com/content/b2b-chemicals/success-strategies-for-indian-pharma-industry-in-an-uncertain-world-114021701557_1.html

PTI. (2015). India's pharma industry may touch $55 Billion by 2020: Assocham and TechSci Report. Retrieved from http://economictimes.indiatimes.com/industry/healthcare/biotech/pharmaceuticals/indias-pharma-industry-may-touch-55-billion-by-2020-assocham-and-techsci-report/articleshow/50353294.cms

Appendix A: Rules, Regulations and Guidelines for Clinical Trials

The rules that apply for the conduct of Clinical Trials as described in the D&C Act are listed

Rule 122A—Application for permission to import new drug
Rule 122B—Application for approval to manufacture new drug other than the drugs classifiable under Schedules C and C(1)
Rule 122D—Permission to import or manufacture fixed dose combination
Rule 122DA—Application for permission to conduct clinical trials for New Drug/Investigational New Drugs
Rule 122DAA—Definition of Clinical trial
Rule 122DAB—Compensation in case of injury or death during clinical trial
Rule 122DAC—Permission to conduct clinical trial
Rule 122DB—Suspension or cancellation of Permission/Approval
Rule 122DD—Registration of Ethics Committee
Rule 122E—Definition of new drug

The 2005 version of Schedule Y has three major sections and 12 appendices as listed below:

Sections:

1. Application for permission:
   (1) Application for permission to import or manufacture new drugs for sale or to undertake clinical trials should be made in Form 44; along with (i) chemical and pharmaceutical information as prescribed in item 2 of Appendix I; (ii) animal pharmacology data as prescribed in item 3 of Appendix I and Appendix IV;
   (2) If the study drug is intended to be imported for the purposes of examination, test or analysis, the application for import of small quantities of drugs for such purpose should also be made in Form 12.
   (3) For drugs indicated in life threatening/serious diseases or diseases of special relevance to the Indian health scenario, the toxicological and clinical data requirements may be abbreviated, deferred or omitted, as deemed appropriate by the Licensing Authority.
2. Clinical trials:
   (1) Approval for clinical trial
   (2) Responsibilities of Sponsor
   (3) Responsibilities of the Investigator(s)
   (4) Informed Consent
   (5) Responsibilities of the Ethics Committee
   (6) Human Pharmacology (phase I)
   (7) Therapeutic exploratory trials (phase II)
   (8) Therapeutic confirmatory trials (phase III)
   (9) Post Marketing Trials (phase IV)
3. Studies in special populations:
   (1) Geriatrics
   (2) Paediatrics
   (3) Pregnant or nursing women
   (4) Post Marketing Surveillance
   (5) Special studies: Bioavailability/Bioequivalence Studies

Schedule Y Appendices:

- Appendix I — Data to be submitted along with the application (Form 44) to conduct clinical trials/import/manufacture of new drugs for marketing in the country
- Appendix IA — Data required to be submitted by an applicant for the grant of permission to import and/or manufacture a new drug already approved in the country
- Appendix II — Structure, contents and format for clinical study reports
- Appendix III — Animal toxicology (Non Clinical toxicity studies)
- Appendix IV — Animal Pharmacology
- Appendix V — Informed Consent

- Appendix VI Fixed Dose Combinations (FDCs)
- Appendix VII Undertaking by the Investigator
- Appendix VIII Ethics Committee
- Appendix IX Stability testing of new drugs
- Appendix X Contents of the proposed protocol for conducting clinical trials
- Appendix XI Data Elements for reporting serious adverse events occurring in a clinical trial
- Appendix XII Compensation in case of injury or death during clinical trial

WORLD MEDICAL ASSOCIATION DECLARATION OF HELSINKI

Ethical Principles for Medical Research Involving Human Subjects

Adopted by the 18th WMA General Assembly, Helsinki, Finland, June 1964, and amended by the: 29th WMA General Assembly, Tokyo, Japan, October 1975

35th WMA General Assembly, Venice, Italy, October 1983 41st WMA General Assembly, Hong Kong, September 1989

48th WMA General Assembly, Somerset West, Republic of South Africa, October 1996 52nd WMA General Assembly, Edinburgh, Scotland, October 2000

53rd WMA General Assembly, Washington 2002 (Note of Clarification on paragraph 29 added) 55th WMA General Assembly, Tokyo 2004 (Note of Clarification on Paragraph 30 added)

59th WMA General Assembly, Seoul, October 2008

A. INTRODUCTION

1. The World Medical Association (WMA) has developed the Declaration of Helsinki as a statement of ethical principles for medical research involving human subjects, including research on identifiable human material and data.

   The Declaration is intended to be read as a whole and each of its constituent paragraphs should not be applied without consideration of all other relevant paragraphs.
2. Although the Declaration is addressed primarily to physicians, the WMA encourages other participants in medical research involving human subjects to adopt these principles.
3. It is the duty of the physician to promote and safeguard the health of patients, including those who are involved in medical research. The physician's knowledge and conscience are dedicated to the fulfilment of this duty.
4. The Declaration of Geneva of the WMA binds the physician with the words, "The health of my patient will be my first consideration," and the International Code of Medical Ethics declares that, "A physician shall act in the patient's best interest when providing medical care."
5. Medical progress is based on research that ultimately must include studies involving human subjects. Populations that are underrepresented in medical research should be provided appropriate access to participation in research.
6. In medical research involving human subjects, the well-being of the individual research subject must take precedence over all other interests.
7. The primary purpose of medical research involving human subjects is to understand the causes, development and effects of diseases and improve preventive, diagnostic and therapeutic interventions (methods, procedures

and treatments). Even the best current interventions must be evaluated continually through research for their safety, effectiveness, efficiency, accessibility and quality.

8. In medical practice and in medical research, most interventions involve risks and burdens.
9. Medical research is subject to ethical standards that promote respect for all human subjects and protect their health and rights. Some research populations are particularly vulnerable and need special protection. These include those who cannot give or refuse consent for themselves and those who may be vulnerable to coercion or undue influence.
10. Physicians should consider the ethical, legal and regulatory norms and standards for research involving human subjects in their own countries as well as applicable international norms and standards. No national or international ethical, legal or regulatory requirement should reduce or eliminate any of the protections for research subjects set forth in this Declaration.

B. PRINCIPLES FOR ALL MEDICAL RESEARCH

11. It is the duty of physicians who participate in medical research to protect the life, health, dignity, integrity, right to self-determination, privacy, and confidentiality of personal information of research subjects.
12. Medical research involving human subjects must conform to generally accepted scientific principles, be based on a thorough knowledge of the scientific literature, other relevant sources of information, and adequate laboratory and, as appropriate, animal experimentation. The welfare of animals used for research must be respected.
13. Appropriate caution must be exercised in the conduct of medical research that may harm the environment.
14. The design and performance of each research study involving human subjects must be clearly described in a research protocol. The protocol should contain a statement of the ethical considerations involved and should indicate how the principles in this Declaration have been addressed. The protocol should include information regarding funding, sponsors, institutional affiliations, other potential conflicts of interest, incentives for subjects and provisions for treating and/or compensating subjects who are harmed as a consequence of participation in the research study. The protocol should describe arrangements for post-study access by study subjects to interventions identified as beneficial in the study or access to other appropriate care or benefits.
15. The research protocol must be submitted for consideration, comment, guidance and approval to a research ethics committee before the study begins. This committee must be independent of the researcher, the sponsor and any other undue influence. It must take into consideration the laws and regulations of the country or countries in which the research is to be performed as well as applicable international norms and standards but

these must not be allowed to reduce or eliminate any of the protections for research subjects set forth in this Declaration. The committee must have the right to monitor ongoing studies. The researcher must provide monitoring information to the committee, especially information about any serious adverse events. No change to the protocol may be made without consideration and approval by the committee.

16. Medical research involving human subjects must be conducted only by individuals with the appropriate scientific training and qualifications. Research on patients or healthy volunteers requires the supervision of a competent and appropriately qualified physician or other health care professional. The responsibility for the protection of research subjects must always rest with the physician or other health care professional and never the research subjects, even though they have given consent.
17. Medical research involving a disadvantaged or vulnerable population or community is only justified if the research is responsive to the health needs and priorities of this population or community and if there is a reasonable likelihood that this population or community stands to benefit from the results of the research.
18. Every medical research study involving human subjects must be preceded by careful assessment of predictable risks and burdens to the individuals and communities involved in the research in comparison with foreseeable benefits to them and to other individuals or communities affected by the condition under investigation.
19. Every clinical trial must be registered in a publicly accessible database before recruitment of the first subject.
20. Physicians may not participate in a research study involving human subjects unless they are confident that the risks involved have been adequately assessed and can be satisfactorily managed. Physicians must immediately stop a study when the risks are found to outweigh the potential benefits or when there is conclusive proof of positive and beneficial results.
21. Medical research involving human subjects may only be conducted if the importance of the objective outweighs the inherent risks and burdens to the research subjects.
22. Participation by competent individuals as subjects in medical research must be voluntary. Although it may be appropriate to consult family members or community leaders, no competent individual may be enrolled in a research study unless he or she freely agrees.
23. Every precaution must be taken to protect the privacy of research subjects and the confidentiality of their personal information and to minimize the impact of the study on their physical, mental and social integrity.
24. In medical research involving competent human subjects, each potential subject must be adequately informed of the aims, methods, sources of funding, any possible conflicts of interest, institutional affiliations of the researcher, the anticipated benefits and potential risks of the study and the discomfort it may entail, and any other relevant aspects of the

study. The potential subject must be informed of the right to refuse to participate in the study or to withdraw consent to participate at any time without reprisal. Special attention should be given to the specific information needs of individual potential subjects as well as to the methods used to deliver the information. After ensuring that the potential subject has understood the information, the physician or another appropriately qualified individual must then seek the potential subject's freely-given informed consent, preferably in writing. If the consent cannot be expressed in writing, the non-written consent must be formally documented and witnessed.

25. For medical research using identifiable human material or data, physicians must normally seek consent for the collection, analysis, storage and/or reuse. There may be situations where consent would be impossible or impractical to obtain for such research or would pose a threat to the validity of the research. In such situations the research may be done only after consideration and approval of a research ethics committee.
26. When seeking informed consent for participation in a research study the physician should be particularly cautious if the potential subject is in a dependent relationship with the physician or may consent under duress. In such situations the informed consent should be sought by an appropriately qualified individual who is completely independent of this relationship.
27. For a potential research subject who is incompetent, the physician must seek informed consent from the legally authorized representative. These individuals must not be included in a research study that has no likelihood of benefit for them unless it is intended to promote the health of the population represented by the potential subject, the research cannot instead be performed with competent persons, and the research entails only minimal risk and minimal burden.
28. When a potential research subject who is deemed incompetent is able to give assent to decisions about participation in research, the physician must seek that assent in addition to the consent of the legally authorized representative. The potential subject's dissent should be respected.
29. Research involving subjects who are physically or mentally incapable of giving consent, for example, unconscious patients, may be done only if the physical or mental condition that prevents giving informed consent is a necessary characteristic of the research population. In such circumstances the physician should seek informed consent from the legally authorized representative. If no such representative is available and if the research cannot be delayed, the study may proceed without informed consent provided that the specific reasons for involving subjects with a condition that renders them unable to give informed consent have been stated in the research protocol and the study has been approved by a research ethics committee. Consent to remain in the research should be obtained as soon as possible from the subject or a legally authorized representative.

30. Authors, editors and publishers all have ethical obligations with regard to the publication of the results of research. Authors have a duty to make publicly available the results of their research on human subjects and are accountable for the completeness and accuracy of their reports. They should adhere to accepted guidelines for ethical reporting. Negative and inconclusive as well as positive results should be published or otherwise made publicly available. Sources of funding, institutional affiliations and conflicts of interest should be declared in the publication. Reports of research not in accordance with the principles of this Declaration should not be accepted for publication.

C. ADDITIONAL PRINCIPLES FOR MEDICAL RESEARCH COMBINED WITH MEDICAL CARE

31. The physician may combine medical research with medical care only to the extent that the research is justified by its potential preventive, diagnostic or therapeutic value and if the physician has good reason to believe that participation in the research study will not adversely affect the health of the patients who serve as research subjects.
32. The benefits, risks, burdens and effectiveness of a new intervention must be tested against those of the best current proven intervention, except in the following circumstances:
    - The use of placebo, or no treatment, is acceptable in studies where no current proven intervention exists; or
    - Where for compelling and scientifically sound methodological reasons the use of placebo is necessary to determine the efficacy or safety of an intervention and the patients who receive placebo or no treatment will not be subject to any risk of serious or irreversible harm. Extreme care must be taken to avoid abuse of this option.
33. At the conclusion of the study, patients entered into the study are entitled to be informed about the outcome of the study and to share any benefits that result from it, for example, access to interventions identified as beneficial in the study or to other appropriate care or benefits.
34. The physician must fully inform the patient which aspects of the care are related to the research. The refusal of a patient to participate in a study or the patient's decision to withdraw from the study must never interfere with the patient-physician relationship.
35. In the treatment of a patient, where proven interventions do not exist or have been ineffective, the physician, after seeking expert advice, with informed consent from the patient or a legally authorized representative, may use an unproven intervention if in the physician's judgement it offers hope of saving life, re-establishing health or alleviating suffering. Where possible, this intervention should be made the object of research, designed to evaluate its safety and efficacy. In all cases, new information should be recorded and, where appropriate, made publicly available.

E6 GOOD CLINICAL PRACTICE: CONSOLIDATED GUIDANCE

INTRODUCTION

Good clinical practice (GCP) is an international ethical and scientific quality standard for designing, conducting, recording, and reporting trials that involve the participation of human subjects. Compliance with this standard provides public assurance that the rights, safety, and well being of trial subjects are protected, consistent with the principles that have their origin in the Declaration of Helsinki, and that the clinical trial data are credible.

The objective of this ICH GCP guidance is to provide a unified standard for the European Union (EU), Japan, and the United States to facilitate the mutual acceptance of clinical data by the regulatory authorities in these jurisdictions. The guidance was developed with consideration of the current good clinical practices of the European Union, Japan, and the United States, as well as those of Australia, Canada, the Nordic countries, and the World Health Organization (WHO).

This guidance should be followed when generating clinical trial data that are intended to be submitted to regulatory authorities. The principles established in this guidance may also be applied to other clinical investigations that may have an impact on the safety and well-being of human subjects.

1. GLOSSARY

1.1 Adverse drug reaction (ADR): In the preapproval clinical experience with a new medicinal product or its new usages, particularly as the therapeutic dose(s) may not be established, all noxious and unintended responses to a medicinal product related to any dose should be considered adverse drug reactions. The phrase "responses to a medicinal product" means that a causal relationship between a medicinal product and an adverse event is at least a reasonable possibility, i.e., the relationship cannot be ruled out. Regarding marketed medicinal products: A response to a drug that is noxious and unintended and that occurs at doses normally used in man for prophylaxis, diagnosis, or therapy of diseases or for modification of physiological function (see the ICH guidance for Clinical Safety Data Management: Definitions and Standards for Expedited Reporting).

1.2 Adverse event (AE): An AE is any untoward medical occurrence in a patient or clinical investigation subject administered a pharmaceutical product and that does not necessarily have a causal relationship with this treatment. An AE can therefore be any unfavorable and unintended sign (including an abnormal laboratory finding), symptom, or disease temporally associated with the use of a medicinal (investigational) product, whether or not related to the medicinal (investigational) product (see the ICH guidance for Clinical Safety Data Management: Definitions and Standards for Expedited Reporting).

1.3 Amendment (to the protocol): See Protocol Amendment.

1.4 **Applicable regulatory requirement(s)**: Any law(s) and regulation(s) addressing the conduct of clinical trials of investigational products of the jurisdiction where trial is conducted.

1.5 **Approval (in relation to institutional review boards (IRBs))**: The affirmative decision of the IRB that the clinical trial has been reviewed and may be conducted at the institution site within the constraints set forth by the IRB, the institution, good clinical practice (GCP), and the applicable regulatory requirements.

1.6 **Audit**: A systematic and independent examination of trial-related activities and documents to determine whether the evaluated trial-related activities were conducted, and the data were recorded, analyzed, and accurately reported according to the protocol, sponsor's standard operating procedures (SOPs), good clinical practice (GCP), and the applicable regulatory requirement(s).

1.7 **Audit certificate**: A declaration of confirmation by the auditor that an audit has taken place.

1.8 **Audit report**: A written evaluation by the sponsor's auditor of the results of the audit.

1.9 **Audit trail**: Documentation that allows reconstruction of the course of events.

1.10 **Blinding/masking**: A procedure in which one or more parties to the trial are kept unaware of the treatment assignment(s). Single blinding usually refers to the subject(s) being unaware, and double blinding usually refers to the subject(s), investigator(s), monitor, and, in some cases, data analyst(s) being unaware of the treatment assignment(s).

1.11 **Case report form (CRF)**: A printed, optical, or electronic document designed to record all of the protocol-required information to be reported to the sponsor on each trial subject.

1.12 **Clinical trial/study**: Any investigation in human subjects intended to discover or verify the clinical, pharmacological, and/or other pharmacodynamic effects of an investigational product(s), and/or to identify any adverse reactions to an investigational product(s), and/or to study absorption, distribution, metabolism, and excretion of an investigational product(s) with the object of ascertaining its safety and/or efficacy. The terms clinical trial and clinical study are synonymous.

1.13 **Clinical Trial/Study Report**: A written description of a trial/study of any therapeutic, prophylactic, or diagnostic agent conducted in human subjects, in which the clinical and statistical description, presentations, and analyses are fully integrated into a single report (see the ICH Guidance for Structure and Content of Clinical Study Reports).

1.14 **Comparator (Product)**: An investigational or marketed product (i.e., active control), or placebo, used as a reference in a clinical trial.

1.15 **Compliance (in relation to trials)**: Adherence to all the trial-related requirements, good clinical practice (GCP) requirements, and the applicable regulatory requirements.

1.16 Confidentiality: Prevention of disclosure, to other than authorized individuals, of a sponsor's proprietary information or of a subject's identity.

1.17 Contract: A written, dated, and signed agreement between two or more involved parties that sets out any arrangements on delegation and distribution of tasks and obligations and, if appropriate, on financial matters. The protocol may serve as the basis of a contract.

1.18 Coordinating Committee: A committee that a sponsor may organize to coordinate the conduct of a multicenter trial.

1.19 Coordinating Investigator: An investigator assigned the responsibility for the coordination of investigators at different centers participating in a multicenter trial.

1.20 Contract Research Organization (CRO): A person or an organization (commercial, academic, or other) contracted by the sponsor to perform one or more of a sponsor's trial-related duties and functions.

1.21 Direct Access: Permission to examine, analyze, verify, and reproduce any records and reports that are important to evaluation of a clinical trial. Any party (for example, domestic and foreign regulatory authorities, sponsors, monitors, and auditors) with direct access should take all reasonable precautions within the constraints of the applicable regulatory requirement(s) to maintain the confidentiality of subjects' identities and sponsor's proprietary information.

1.22 Documentation: All records, in any form (including, but not limited to, written, electronic, magnetic, and optical records; and scans, x-rays, and electrocardiograms) that describe or record the methods, conduct, and/or results of a trial, the factors affecting a trial, and the actions taken.

1.23 Essential Documents: Documents that individually and collectively permit evaluation of the conduct of a study and the quality of the data produced (see Section 8. "Essential Documents for the Conduct of a Clinical Trial").

1.24 Good Clinical Practice (GCP): A standard for the design, conduct, performance, monitoring, auditing, recording, analyses, and reporting of clinical trials that provides assurance that the data and reported results are credible and accurate, and that the rights, integrity, and confidentiality of trial subjects are protected.

1.25 Independent Data Monitoring Committee (IDMC) (Data and Safety Monitoring Board, Monitoring Committee, Data Monitoring Committee): An independent data monitoring committee that may be established by the sponsor to assess at intervals the progress of a clinical trial, the safety data, and the critical efficacy endpoints, and to recommend to the sponsor whether to continue, modify, or stop a trial.

1.26 Impartial Witness: A person, who is independent of the trial, who cannot be unfairly influenced by people involved with the trial, who attends the informed consent process if the subject or the subject's legally acceptable representative cannot read, and who reads the informed consent form and any other written information supplied to the subject.

1.27 Independent Ethics Committee (IEC): An independent body (a review board or a committee, institutional, regional, national, or supranational), constituted of medical/scientific professionals and nonmedical/nonscientific members, whose responsibility it is to ensure the protection of the rights, safety, and well-being of human subjects involved in a trial and to provide public assurance of that protection, by, among other things, reviewing and approving/providing favorable opinion on the trial protocol, the suitability of the investigator(s), facilities, and the methods and material to be used in obtaining and documenting informed consent of the trial subjects. The legal status, composition, function, operations, and regulatory requirements pertaining to Independent Ethics Committees may differ among countries, but should allow the Independent Ethics Committee to act in agreement with GCP as described in this guidance.

1.28 Informed Consent: A process by which a subject voluntarily confirms his or her willingness to participate in a particular trial, after having been informed of all aspects of the trial that are relevant to the subject's decision to participate. Informed consent is documented by means of a written, signed, and dated informed consent form.

1.29 Inspection: The act by a regulatory authority(ies) of conducting an official review of documents, facilities, records, and any other resources that are deemed by the authority(ies) to be related to the clinical trial and that may be located at the site of the trial, at the sponsor's and/or contract research organization's (CROs) facilities, or at other establishments deemed appropriate by the regulatory authority(ies).

1.30 Institution (medical): Any public or private entity or agency or medical or dental facility where clinical trials are conducted.

1.31 Institutional Review Board (IRB): An independent body constituted of medical, scientific, and nonscientific members, whose responsibility it is to ensure the protection of the rights, safety, and well-being of human subjects involved in a trial by, among other things, reviewing, approving, and providing continuing review of trials, of protocols and amendments, and of the methods and material to be used in obtaining and documenting informed consent of the trial subjects.

1.32 Interim Clinical Trial/Study Report: A report of intermediate results and their evaluation based on analyses performed during the course of a trial.

1.33 Investigational Product: A pharmaceutical form of an active ingredient or placebo being tested or used as a reference in a clinical trial, including a product with a marketing authorization when used or assembled (formulated or packaged) in a way different from the approved form, or when used for an unapproved indication, or when used to gain further information about an approved use.

1.34 Investigator: A person responsible for the conduct of the clinical trial at a trial site. If a trial is conducted by a team of individuals at a trial site, the investigator is the responsible leader of the team and may be called the principal investigator. See also Subinvestigator.

1.35 Investigator/Institution: An expression meaning "the investigator and/or institution, where required by the applicable regulatory requirements."

1.36 Investigator's Brochure: A compilation of the clinical and nonclinical data on the investigational product(s) that is relevant to the study of the investigational product(s) in human subjects (see Section 7. "Investigator's Brochure").

1.37 Legally Acceptable Representative: An individual or juridical or other body authorized under applicable law to consent, on behalf of a prospective subject, to the subject's participation in the clinical trial.

1.38 Monitoring: The act of overseeing the progress of a clinical trial, and of ensuring that it is conducted, recorded, and reported in accordance with the protocol, standard operating procedures (SOPs), GCP, and the applicable regulatory requirement(s).

1.39 Monitoring Report: A written report from the monitor to the sponsor after each site visit and/or other trial-related communication according to the sponsor's SOPs.

1.40 Multicenter Trial: A clinical trial conducted according to a single protocol but at more than one site, and, therefore, carried out by more than one investigator.

1.41 Nonclinical Study: Biomedical studies not performed on human subjects.

1.42 Opinion (in relation to Independent Ethics Committee): The judgment and/or the advice provided by an Independent Ethics Committee (IEC).

1.43 Original Medical Record: See Source Documents.

1.44 Protocol: A document that describes the objective(s), design, methodology, statistical considerations, and organization of a trial. The protocol usually also gives the background and rationale for the trial, but these could be provided in other protocol referenced documents. Throughout the ICH GCP Guidance, the term protocol refers to protocol and protocol amendments.

1.45 Protocol Amendment: A written description of a change(s) to or formal clarification of a protocol.

1.46 Quality Assurance (QA): All those planned and systematic actions that are established to ensure that the trial is performed and the data are generated, documented (recorded), and reported in compliance with GCP and the applicable regulatory requirement(s).

1.47 Quality Control (QC): The operational techniques and activities undertaken within the quality assurance system to verify that the requirements for quality of the trial related activities have been fulfilled.

1.48 Randomization: The process of assigning trial subjects to treatment or control groups using an element of chance to determine the assignments in order to reduce bias.

1.49 Regulatory Authorities: Bodies having the power to regulate. In the ICH GCP guidance, the expression "Regulatory Authorities" includes the authorities that review submitted clinical data and those that conduct inspections (see Section 1.29). These bodies are sometimes referred to as competent authorities.

1.50 Serious Adverse Event (SAE) or Serious Adverse Drug Reaction (Serious ADR): Any untoward medical occurrence that at any dose:

- Results in death,
- Is life-threatening,
- Requires inpatient hospitalization or prolongation of existing hospitalization,
- Results in persistent or significant disability/incapacity, or
- Is a congenital anomaly/birth defect.

(See the ICH guidance for Clinical Safety Data Management: Definitions and Standards for Expedited Reporting.)

1.51 Source Data: All information in original records and certified copies of original records of clinical findings, observations, or other activities in a clinical trial necessary for the reconstruction and evaluation of the trial. Source data are contained in source documents (original records or certified copies).

1.52 Source Documents: Original documents, data, and records (for example, hospital records, clinical and office charts, laboratory notes, memoranda, subjects' diaries or evaluation checklists, pharmacy dispensing records, recorded data from automated instruments, copies or transcriptions certified after verification as being accurate and complete, microfiches, photographic negatives, microfilm or magnetic media, x-rays, subject files, and records kept at the pharmacy, at the laboratories, and at medico-technical departments involved in the clinical trial).

1.53 Sponsor: An individual, company, institution, or organization that takes responsibility for the initiation, management, and/or financing of a clinical trial.

1.54 Sponsor-Investigator: An individual who both initiates and conducts, alone or with others, a clinical trial, and under whose immediate direction the investigational product is administered to, dispensed to, or used by a subject. The term does not include any person other than an individual (for example, it does not include a corporation or an agency). The obligations of a sponsor-investigator include both those of a sponsor and those of an investigator.

1.55 Standard Operating Procedures (SOPs): Detailed, written instructions to achieve uniformity of the performance of a specific function.

1.56 Subinvestigator: Any individual member of the clinical trial team designated and supervised by the investigator at a trial site to perform critical trial-related procedures and/or to make important trial-related decisions (for example, associates, residents, research fellows). See also Investigator.

1.57 Subject/Trial Subject: An individual who participates in a clinical trial, either as a recipient of the investigational product(s) or as a control.

1.58 Subject Identification Code: A unique identifier assigned by the investigator to each trial subject to protect the subject's identity and used in lieu of the subject's name when the investigator reports adverse events and/or other trial-related data.

1.59 Trial Site: The location(s) where trial-related activities are actually conducted.

1.60 Unexpected Adverse Drug Reaction: An adverse reaction, the nature or severity of which is not consistent with the applicable product information (for example, Investigator's Brochure for an unapproved investigational product or package insert/summary of product characteristics for an approved product). (See the ICH Guidance for Clinical Safety Data Management: Definitions and Standards for Expedited Reporting.)

1.61 Vulnerable Subjects: Individuals whose willingness to volunteer in a clinical trial may be unduly influenced by the expectation, whether justified or not, of benefits associated with participation, or of a retaliatory response from senior members of a hierarchy in case of refusal to participate. Examples are members of a group with a hierarchical structure, such as medical, pharmacy, dental, and nursing students, subordinate hospital and laboratory personnel, employees of the pharmaceutical industry, members of the armed forces, and persons kept in detention. Other vulnerable subjects include patients with incurable diseases, persons in nursing homes, unemployed or impoverished persons, patients in emergency situations, ethnic minority groups, homeless persons, nomads, refugees, minors, and those incapable of giving consent.

1.62 Well-being (of the trial subjects): The physical and mental integrity of the subjects participating in a clinical trial.

2. THE PRINCIPLES OF ICH GCP

2.1 Clinical trials should be conducted in accordance with the ethical principles that have their origin in the Declaration of Helsinki, and that are consistent with GCP and the applicable regulatory requirement(s).

2.2 Before a trial is initiated, foreseeable risks and inconveniences should be weighed against the anticipated benefit for the individual trial subject and society. A trial should be initiated and continued only if the anticipated benefits justify the risks.

2.3 The rights, safety, and well-being of the trial subjects are the most important considerations and should prevail over interests of science and society.

2.4 The available nonclinical and clinical information on an investigational product should be adequate to support the proposed clinical trial.

2.5 Clinical trials should be scientifically sound, and described in a clear, detailed protocol.

2.6 A trial should be conducted in compliance with the protocol that has received prior institutional review board (IRB)/independent ethics committee (IEC) approval/favorable opinion.

2.7 The medical care given to, and medical decisions made on behalf of, subjects should always be the responsibility of a qualified physician or, when appropriate, of a qualified dentist.

2.8 Each individual involved in conducting a trial should be qualified by education, training, and experience to perform his or her respective task(s).

2.9 Freely given informed consent should be obtained from every subject prior to clinical trial participation.

2.10 All clinical trial information should be recorded, handled, and stored in a way that allows its accurate reporting, interpretation, and verification.

2.11 The confidentiality of records that could identify subjects should be protected, respecting the privacy and confidentiality rules in accordance with the applicable regulatory requirement(s).

2.12 Investigational products should be manufactured, handled, and stored in accordance with applicable good manufacturing practice (GMP). They should be used in accordance with the approved protocol.

2.13 Systems with procedures that assure the quality of every aspect of the trial should be implemented.

3. INSTITUTIONAL REVIEW BOARD/INDEPENDENT ETHICS COMMITTEE (IRB/IEC)

3.1 Responsibilities

3.1.1 An IRB/IEC should safeguard the rights, safety, and well-being of all trial subjects. Special attention should be paid to trials that may include vulnerable subjects.

3.1.2 The IRB/IEC should obtain the following documents: Trial protocol(s)/amendment(s), written informed consent form(s) and consent form updates that the investigator proposes for use in the trial, subject recruitment procedures (for example, advertisements), written information to be provided to subjects, Investigator's Brochure (IB), available safety information, information about payments and compensation available to subjects, the investigator's current curriculum vitae and/or other documentation evidencing qualifications, and any other documents that the IRB/IEC may require to fulfil its responsibilities. The IRB/IEC should review a proposed clinical trial within a reasonable time and document its views in writing, clearly identifying the trial, the documents reviewed, and the dates for the following:

- Approval/favorable opinion;
- Modifications required prior to its approval/favorable opinion;
- Disapproval/negative opinion; and
- Termination/suspension of any prior approval/favorable opinion.

3.1.3 The IRB/IEC should consider the qualifications of the investigator for the proposed trial, as documented by a current curriculum vitae and/or by any other relevant documentation the IRB/IEC requests.

3.1.4 The IRB/IEC should conduct continuing review of each ongoing trial at intervals appropriate to the degree of risk to human subjects, but at least once per year.

3.1.5 The IRB/IEC may request more information than is outlined in paragraph 4.8.10 be given to subjects when, in the judgment of the IRB/IEC, the additional information would add meaningfully to the protection of the rights, safety, and/or well-being of the subjects.

3.1.6 When a nontherapeutic trial is to be carried out with the consent of the subject's legally acceptable representative (see Sections 4.8.12 and 4.8.14), the IRB/IEC should determine that the proposed protocol and/or other document(s) adequately addresses relevant ethical concerns and meets applicable regulatory requirements for such trials.

3.1.7 Where the protocol indicates that prior consent of the trial subject or the subject's legally acceptable representative is not possible (see Section 4.8.15), the IRB/IEC should determine that the proposed protocol and/or other document(s) adequately addresses relevant ethical concerns and meets applicable regulatory requirements for such trials (i.e., in emergency situations).

3.1.8 The IRB/IEC should review both the amount and method of payment to subjects to assure that neither presents problems of coercion or undue influence on the trial subjects. Payments to a subject should be prorated and not wholly contingent on completion of the trial by the subject.

3.1.9 The IRB/IEC should ensure that information regarding payment to subjects, including the methods, amounts, and schedule of payment to trial subjects, is set forth in the written informed consent form and any other written information to be provided to subjects. The way payment will be prorated should be specified.

3.2 Composition, Functions, and Operations

3.2.1 The IRB/IEC should consist of a reasonable number of members, who collectively have the qualifications and experience to review and evaluate the science, medical aspects, and ethics of the proposed trial. It is recommended that the IRB/IEC should include:

(a) At least five members.

(b) At least one member whose primary area of interest is in a nonscientific area.

(c) At least one member who is independent of the institution/trial site.

Only those IRB/IEC members who are independent of the investigator and the sponsor of the trial should vote/provide opinion on a trial-related matter. A list of IRB/IEC members and their qualifications should be maintained.

3.2.2 The IRB/IEC should perform its functions according to written operating procedures, should maintain written records of its activities and minutes of its meetings, and should comply with GCP and with the applicable regulatory requirement(s).

3.2.3 An IRB/IEC should make its decisions at announced meetings at which at least a quorum, as stipulated in its written operating procedures, is present.

3.2.4 Only members who participate in the IRB/IEC review and discussion should vote/provide their opinion and/or advise.

3.2.5 The investigator may provide information on any aspect of the trial, but should not participate in the deliberations of the IRB/IEC or in the vote/opinion of the IRB/IEC.

3.2.6 An IRB/IEC may invite nonmembers with expertise in special areas for assistance.

3.3 Procedures

The IRB/IEC should establish, document in writing, and follow its procedures, which should include:

3.3.1 Determining its composition (names and qualifications of the members) and the authority under which it is established.

3.3.2 Scheduling, notifying its members of, and conducting its meetings.

3.3.3 Conducting initial and continuing review of trials.

3.3.4 Determining the frequency of continuing review, as appropriate.

3.3.5 Providing, according to the applicable regulatory requirements, expedited review and approval/favorable opinion of minor change(s) in ongoing trials that have the approval/favorable opinion of the IRB/IEC.

3.3.6 Specifying that no subject should be admitted to a trial before the IRB/IEC issues its written approval/favorable opinion of the trial.

3.3.7 Specifying that no deviations from, or changes of, the protocol should be initiated without prior written IRB/IEC approval/favorable opinion of an appropriate amendment, except when necessary to eliminate immediate hazards to the subjects or when the change(s) involves only logistical or administrative aspects of the trial (for example, change of monitor(s), telephone number(s)) (see Section 4.5.2).

3.3.8 Specifying that the investigator should promptly report to the IRB/IEC:

(a) Deviations from, or changes of, the protocol to eliminate immediate hazards to the trial subjects (see Sections 3.3.7, 4.5.2, and 4.5.4).

(b) Changes increasing the risk to subjects and/or affecting significantly the conduct of the trial (see Section 4.10.2).

(c) All adverse drug reactions (ADRs) that are both serious and unexpected.
(d) New information that may affect adversely the safety of the subjects or the conduct of the trial.

3.3.9 Ensuring that the IRB/IEC promptly notify in writing the investigator/institution concerning:
(a) Its trial-related decisions/opinions.
(b) The reasons for its decisions/opinions.
(c) Procedures for appeal of its decisions/opinions.

3.4 Records

The IRB/IEC should retain all relevant records (for example, written procedures, membership lists, lists of occupations/affiliations of members, submitted documents, minutes of meetings, and correspondence) for a period of at least 3 years after completion of the trial and make them available upon request from the regulatory authority(ies). The IRB/IEC may be asked by investigators, sponsors, or regulatory authorities to provide copies of its written procedures and membership lists.

4. INVESTIGATOR

4.1 Investigator's Qualifications and Agreements

4.1.1 The investigator(s) should be qualified by education, training, and experience to assume responsibility for the proper conduct of the trial, should meet all the qualifications specified by the applicable regulatory requirement(s), and should provide evidence of such qualifications through up-to-date curriculum vitae and/or other relevant documentation requested by the sponsor, the IRB/IEC, and/or the regulatory authority(ies).

4.1.2 The investigator should be thoroughly familiar with the appropriate use of the investigational product(s), as described in the protocol, in the current Investigator's Brochure, in the product information, and in other information sources provided by the sponsor.

4.1.3 The investigator should be aware of, and should comply with, GCP and the applicable regulatory requirements.

4.1.4 The investigator/institution should permit monitoring and auditing by the sponsor, and inspection by the appropriate regulatory authority(ies).

4.1.5 The investigator should maintain a list of appropriately qualified persons to whom the investigator has delegated significant trial-related duties.

4.2 Adequate Resources

4.2.1 The investigator should be able to demonstrate (for example, based on retrospective data) a potential for recruiting the required number of suitable subjects within the agreed recruitment period.

4.2.2 The investigator should have sufficient time to properly conduct and complete the trial within the agreed trial period.

4.2.3 The investigator should have available an adequate number of qualified staff and adequate facilities for the foreseen duration of the trial to conduct the trial properly and safely.

4.2.4 The investigator should ensure that all persons assisting with the trial are adequately informed about the protocol, the investigational product(s), and their trial-related duties and functions.

4.3 Medical Care of Trial Subjects

4.3.1 A qualified physician (or dentist, when appropriate), who is an investigator or a subinvestigator for the trial, should be responsible for all trial-related medical (or dental) decisions.

4.3.2 During and following a subject's participation in a trial, the investigator/institution should ensure that adequate medical care is provided to a subject for any adverse events, including clinically significant laboratory values, related to the trial. The investigator/institution should inform a subject when medical care is needed for intercurrent illness(es) of which the investigator becomes aware.

4.3.3 It is recommended that the investigator inform the subject's primary physician about the subject's participation in the trial if the subject has a primary physician and if the subject agrees to the primary physician being informed.

4.3.4 Although a subject is not obliged to give his/her reason(s) for withdrawing prematurely from a trial, the investigator should make a reasonable effort to ascertain the reason(s), while fully respecting the subject's rights.

4.4 Communication with IRB/IEC

4.4.1 Before initiating a trial, the investigator/institution should have written and dated approval/favorable opinion from the IRB/IEC for the trial protocol, written informed consent form, consent form updates, subject recruitment procedures (for example, advertisements), and any other written information to be provided to subjects.

4.4.2 As part of the investigator's/institution's written application to the IRB/IEC, the investigator/institution should provide the IRB/IEC with a current copy of the Investigator's Brochure. If the Investigator's Brochure is updated during the trial, the investigator/institution should supply a copy of the updated Investigator's Brochure to the IRB/IEC.

4.4.3 During the trial the investigator/institution should provide to the IRB/IEC all documents subject to its review.

4.5 Compliance with Protocol

4.5.1 The investigator/institution should conduct the trial in compliance with the protocol agreed to by the sponsor and, if required, by the regulatory authority(ies), and which was given approval/favorable opinion by the IRB/IEC. The investigator/institution and the sponsor should sign the protocol, or an alternative contract, to confirm their agreement.

4.5.2 The investigator should not implement any deviation from, or changes of, the protocol without agreement by the sponsor and prior review and documented approval/favorable opinion from the IRB/IEC of an amendment, except where necessary to eliminate an immediate hazard(s) to trial subjects, or when the change(s) involves only logistical or administrative aspects of the trial (for example, change of monitor(s), change of telephone number(s)).

4.5.3 The investigator, or person designated by the investigator, should document and explain any deviation from the approved protocol.

4.5.4 The investigator may implement a deviation from, or a change in, the protocol to eliminate an immediate hazard(s) to trial subjects without prior IRB/IEC approval/favorable opinion. As soon as possible, the implemented deviation or change, the reasons for it, and, if appropriate, the proposed protocol amendment(s) should be submitted:

(a) To the IRB/IEC for review and approval/favorable opinion;

(b) To the sponsor for agreement and, if required;

(c) To the regulatory authority(ies).

4.6 Investigational Product(s)

4.6.1 Responsibility for investigational product(s) accountability at the trial site(s)

4.6.2 Where allowed/required, the investigator/institution may/should assign some or all of the investigator's/institution's duties for investigational product(s) accountability at the trial site(s) to an appropriate pharmacist or another appropriate individual who is under the supervision of the investigator/institution.

4.6.3 The investigator/institution and/or a pharmacist or other appropriate individual, who is designated by the investigator/institution, should maintain records of the product's delivery to the trial site, the inventory at the site, the use by each subject, and the return to the sponsor or alternative disposition of unused product(s). These records should include dates, quantities, batch/serial numbers, expiration dates (if applicable), and the unique code numbers assigned to the investigational product(s) and trial subjects. Investigators should maintain records that document adequately that the subjects were provided the doses specified by the protocol and reconcile all investigational product(s) received from the sponsor.

4.6.4 The investigational product(s) should be stored as specified by the sponsor (see Sections 5.13.2 and 5.14.3) and in accordance with applicable regulatory requirement(s).

4.6.5 The investigator should ensure that the investigational product(s) are used only in accordance with the approved protocol.

4.6.6 The investigator, or a person designated by the investigator/institution, should explain the correct use of the investigational product(s) to each subject and should check, at intervals appropriate for the trial, that each subject is following the instructions properly.

4.7 Randomization Procedures and Unblinding

The investigator should follow the trial's randomization procedures, if any, and should ensure that the code is broken only in accordance with the protocol. If the trial is blinded, the investigator should promptly document and explain to the sponsor any premature unblinding (for example, accidental unblinding, unblinding due to a serious adverse event) of the investigational product(s).

4.8 Informed Consent of Trial Subjects

4.8.1 In obtaining and documenting informed consent, the investigator should comply with the applicable regulatory requirement(s), and should adhere to GCP and to the ethical principles that have their origin in the Declaration of Helsinki. Prior to the beginning of the trial, the investigator should have the IRB/IEC's written approval/favorable opinion of the written informed consent form and any other written information to be provided to subjects.

4.8.2 The written informed consent form and any other written information to be provided to subjects should be revised whenever important new information becomes available that may be relevant to the subject's consent. Any revised written informed consent form, and written information should receive the IRB/IEC's approval/favorable opinion in advance of use. The subject or the subject's legally acceptable representative should be informed in a timely manner if new information becomes available that may be relevant to the subject's willingness to continue participation in the trial. The communication of this information should be documented.

4.8.3 Neither the investigator, nor the trial staff, should coerce or unduly influence a subject to participate or to continue to participate in a trial.

4.8.4 None of the oral and written information concerning the trial, including the written informed consent form, should contain any language that causes the subject or the subject's legally acceptable representative to waive or to appear to waive any legal rights, or that releases or appears to release the investigator, the institution, the sponsor, or their agents from liability for negligence.

4.8.5 The investigator, or a person designated by the investigator, should fully inform the subject or, if the subject is unable to provide informed consent, the subject's legally acceptable representative, of all pertinent aspects of the trial including the written information given approval/favorable opinion by the IRB/IEC.

4.8.6 The language used in the oral and written information about the trial, including the written informed consent form, should be as nontechnical as practical and should be understandable to the subject or the subject's legally acceptable representative and the impartial witness, where applicable.

4.8.7 Before informed consent may be obtained, the investigator, or a person designated by the investigator, should provide the subject or the subject's legally acceptable representative ample time and opportunity to inquire about details of the trial and to decide whether or not to participate in the trial. All questions about the trial should be answered to the satisfaction of the subject or the subject's legally acceptable representative.

4.8.8 Prior to a subject's participation in the trial, the written informed consent form should be signed and personally dated by the subject or by the subject's legally acceptable representative, and by the person who conducted the informed consent discussion.

4.8.9 If a subject is unable to read or if a legally acceptable representative is unable to read, an impartial witness should be present during the entire informed consent discussion. After the written informed consent form and any other written information to be provided to subjects is read and explained to the subject or the subject's legally acceptable representative, and after the subject or the subject's legally acceptable representative has orally consented to the subject's participation in the trial, and, if capable of doing so, has signed and personally dated the informed consent form, the witness should sign and personally date the consent form. By signing the consent form, the witness attests that the information in the representative, and that informed consent was freely given by the subject or the subject's legally acceptable representative.

4.8.10 Both the informed consent discussion and the written informed consent

(a) The purpose of the trial.
(b) The trial treatment(s) and the probability for random assignment to each treatment.
(c) The trial procedures to be followed, including all invasive procedures.
(d) The subject's responsibilities.
(e) Those aspects of the trial that are experimental.
(f) The reasonably foreseeable risks or inconveniences to the subject and, when applicable, to an embryo, fetus, or nursing infant.
(g) The reasonably expected benefits. When there is no intended clinical benefit to the subject, the subject should be made aware of this.

(h) The alternative procedure(s) or course(s) of treatment that may be available to the subject, and their important potential benefits and risks.
(i) The compensation and/or treatment available to the subject in the event of trial-related injury.
(j) The anticipated prorated payment, if any, to the subject for participating in the trial.
(k) The anticipated expenses, if any, to the subject for participating in the trial.
(l) That the subject's participation in the trial is voluntary and that the subject may refuse to participate or withdraw from the trial, at any time, without penalty or loss of benefits to which the subject is otherwise entitled.
(m) That the monitor(s), the auditor(s), the IRB/IEC, and the regulatory authority(ies) will be granted direct access to the subject's original medical records for verification of clinical trial procedures and/or data, without violating the confidentiality of the subject, to the extent permitted by the applicable laws and regulations and that, by signing a written informed consent form, the subject or the subject's legally acceptable representative is authorizing such access.
(n) That records identifying the subject will be kept confidential and, to the extent permitted by the applicable laws and/or regulations, will not be made publicly available. If the results of the trial are published, the subject's identity will remain confidential.
(o) That the subject or the subject's legally acceptable representative will be informed in a timely manner if information becomes available that may be relevant to the subject's willingness to continue participation in the trial.
(p) The person(s) to contact for further information regarding the trial and the rights of trial subjects, and whom to contact in the event of trial-related injury.
(q) The foreseeable circumstances and/or reasons under which the subject's participation in the trial may be terminated.
(r) The expected duration of the subject's participation in the trial.
(s) The approximate number of subjects involved in the trial.

4.8.11 Prior to participation in the trial, the subject or the subject's legally acceptable representative should receive a copy of the signed and dated written informed consent form and any other written information provided to the subjects. During a subject's participation in the trial, the subject or the subject's legally acceptable representative should receive a copy of the signed and dated consent form updates and a copy of any amendments to the written information provided to subjects.

4.8.12 When a clinical trial (therapeutic or nontherapeutic) includes subjects who can only be enrolled in the trial with the consent of the subject's legally acceptable representative (for example, minors, or patients with severe dementia), the subject should be informed about the trial to the extent compatible with the subject's understanding and, if capable, the subject should assent, sign and personally date the written informed consent.

4.8.13 Except as described in 4.8.14, a nontherapeutic trial (i.e., a trial in which there is no anticipated direct clinical benefit to the subject) should be conducted in subjects who personally give consent and who sign and date the written informed consent form.

4.8.14 Nontherapeutic trials may be conducted in subjects with consent of a legally acceptable representative provided the following conditions are fulfilled:

(a) The objectives of the trial cannot be met by means of a trial in subjects who can give informed consent personally.
(b) The foreseeable risks to the subjects are low.
(c) The negative impact on the subject's well-being is minimized and low.
(d) The trial is not prohibited by law.
(e) The approval/favorable opinion of the IRB/IEC is expressly sought on the inclusion of such subjects, and the written approval/favorable opinion covers this aspect. Such trials, unless an exception is justified, should be conducted in patients having a disease or condition for which the investigational product is intended. Subjects in these trials should be particularly closely monitored and should be withdrawn if they appear to be unduly distressed.

4.8.15 In emergency situations, when prior consent of the subject is not possible, the consent of the subject's legally acceptable representative, if present, should be requested. When prior consent of the subject is not possible, and the subject's legally acceptable representative is not available, enrollment of the subject should require measures described in the protocol and/or elsewhere, with documented approval/favorable opinion by the IRB/IEC, to protect the rights, safety, and wellbeing of the subject and to ensure compliance with applicable regulatory requirements. The subject or the subject's legally acceptable representative should be informed about the trial as soon as possible and consent to continue and other consent as appropriate (see Section 4.8.10) should be requested.

4.9 Records and Reports

4.9.1 The investigator should ensure the accuracy, completeness, legibility, and timeliness of the data reported to the sponsor in the CRFs and in all required.

4.9.2 Data reported on the CRF, which are derived from source documents, should be consistent with the source documents or the discrepancies should be explained.

4.9.3 Any change or correction to a CRF should be dated, initialed, and explained (if necessary) and should not obscure the original entry (i.e., an audit trail should be maintained); this applies to both written and electronic changes or corrections (see Section 5.18.4(n)). Sponsors should provide guidance to investigators and/or the investigators' designated representatives on making such corrections. Sponsors should have written procedures to assure that changes or corrections in CRFs made by sponsor's designated representatives are documented, are necessary, and are endorsed by the investigator. The investigator should retain records of the changes and corrections.

4.9.4 The investigator/institution should maintain the trial documents as specified in Essential Documents for the Conduct of a Clinical Trial (see Section 8) and as required by the applicable regulatory requirement(s). The investigator/institution should take measures to prevent accidental or premature destruction of these documents.

4.9.5 Essential documents should be retained until at least 2 years after the last approval of a marketing application in an ICH region and until there are no pending or contemplated marketing applications in an ICH region or at least 2 years have elapsed since the formal discontinuation of clinical development of the investigational product. These documents should be retained for a longer period, however, if required by the applicable regulatory requirements or by an agreement with the sponsor. It is the responsibility of the sponsor to inform the investigator/institution as to when these documents no longer need to be retained (see Section 5.5.12).

4.9.6 The financial aspects of the trial should be documented in an agreement between the sponsor and the investigator/institution.

4.9.7 Upon request of the monitor, auditor, IRB/IEC, or regulatory authority, the investigator/institution should make available for direct access all requested trial related records.

4.10 Progress Reports

4.10.1 Where required by the applicable regulatory requirements, the investigator should submit written summaries of the trial's status to the institution. The investigator/institution should submit written summaries of the status of the trial to the IRB/IEC annually, or more frequently, if requested by the IRB/IEC.

4.10.2 The investigator should promptly provide written reports to the sponsor, the IRB/IEC (see Section 3.3.8), and, where required by the applicable regulatory requirements, the institution on any changes significantly affecting the conduct of the trial, and/or increasing the risk to subjects.

4.11 Safety Reporting

4.11.1 All serious adverse events (SAEs) should be reported immediately to the sponsor except for those SAEs that the protocol or other document (for example, Investigator's Brochure) identifies as not needing immediate reporting. The immediate reports should be followed promptly by detailed, written reports. The immediate and follow-up reports should identify subjects by unique code numbers assigned to the trial subjects rather than by the subjects' names, personal identification numbers, and/or addresses. The investigator should also comply with the applicable regulatory requirement(s) related to the reporting of unexpected serious adverse drug reactions to the regulatory authority(ies) and the IRB/IEC.

4.11.2 Adverse events and/or laboratory abnormalities identified in the protocol as critical to safety evaluations should be reported to the sponsor according to the reporting requirements and within the time periods specified by the sponsor in the protocol.

4.11.3 For reported deaths, the investigator should supply the sponsor and the IRB/IEC with any additional requested information (for example, autopsy reports and terminal medical reports).

4.12 Premature Termination or Suspension of a Trial

If the trial is terminated prematurely or suspended for any reason, the investigator/institution should promptly inform the trial subjects, should assure appropriate therapy and follow-up for the subjects, and, where required by the applicable regulatory requirement(s), should inform the regulatory authority(ies). In addition:

4.12.1 If the investigator terminates or suspends a trial without prior agreement of the sponsor, the investigator should inform the institution, where required by the applicable regulatory requirements, and the investigator/institution should promptly inform the sponsor and the IRB/IEC, and should provide the sponsor and the IRB/IEC a detailed written explanation of the termination or suspension.

4.12.2 If the sponsor terminates or suspends a trial (see Section 5.21), the applicable regulatory requirements, and the investigator/institution should promptly inform the IRB/IEC and provide the IRB/IEC a detailed written explanation of the termination or suspension.

4.12.3 If the IRB/IEC terminates or suspends its approval/favorable opinion of a trial (see Sections 3.1.2 and 3.3.9), the investigator should inform the institution, where required by the applicable regulatory requirements, and the investigator/institution should promptly notify the sponsor and provide the sponsor with a detailed written explanation of the termination or suspension.

4.13 Final Report(s) by Investigator/Institution

Upon completion of the trial, the investigator should, where required by the applicable regulatory requirements, inform the institution, and the investigator/institution should provide the sponsor with all required reports, the IRB/IEC with a summary of the trial's outcome, and the regulatory authority(ies) with any report(s) they require of the investigator/institution.

5. SPONSOR

5.1 Quality Assurance and Quality Control

5.1.1 The sponsor is responsible for implementing and maintaining quality assurance and quality control systems with written SOPs to ensure that trials are conducted and data are generated, documented (recorded), and reported in compliance with the protocol, GCP, and the applicable regulatory requirement(s).

5.1.2 The sponsor is responsible for securing agreement from all involved parties to ensure direct access (see Section 1.21) to all trial-related sites, source data/documents, and reports for the purpose of monitoring and auditing by the sponsor, and inspection by domestic and foreign regulatory authorities.

5.1.3 Quality control should be applied to each stage of data handling to ensure that all data are reliable and have been processed correctly.

5.1.4 Agreements, made by the sponsor with the investigator/institution and/or with any other parties involved with the clinical trial, should be in writing, as part of the protocol or in a separate agreement.

5.2 Contract Research Organization (CRO)

5.2.1 A sponsor may transfer any or all of the sponsor's trial-related duties and functions to a CRO, but the ultimate responsibility for the quality and integrity of the trial data always resides with the sponsor. The CRO should implement quality assurance and quality control.

5.2.2 Any trial-related duty and function that is transferred to and assumed by a CRO should be specified in writing.

5.2.3 Any trial-related duties and functions not specifically transferred to and assumed by a CRO are retained by the sponsor.

5.2.4 All references to a sponsor in this guidance also apply to a CRO to the extent that a CRO has assumed the trial-related duties and functions of a sponsor.

5.3 Medical Expertise

The sponsor should designate appropriately qualified medical personnel who will be readily available to advise on trial-related medical questions or problems. If necessary, outside consultant(s) may be appointed for this purpose.

5.4 Trial Design

5.4.1 The sponsor should utilize qualified individuals (for example, biostatisticians, clinical pharmacologists, and physicians) as appropriate, throughout all stages of the trial process, from designing the protocol and CRFs and planning the analyses to analyzing and preparing interim and final clinical trial/study reports.

5.4.2 For further guidance: Clinical Trial Protocol and Protocol Amendment(s) (see Section 6), the ICH Guidance for Structure and Content of Clinical Study Reports, and other appropriate ICH guidance on trial design, protocol, and conduct.

5.5 Trial Management, Data Handling, Recordkeeping, and Independent Data Monitoring Committee

5.5.1 The sponsor should utilize appropriately qualified individuals to supervise the overall conduct of the trial, to handle the data, to verify the data, to conduct the statistical analyses, and to prepare the trial reports.

5.5.2 The sponsor may consider establishing an independent data monitoring committee (IDMC) to assess the progress of a clinical trial, including the safety data and the critical efficacy endpoints at intervals, and to recommend to the sponsor whether to continue, modify, or stop a trial. The IDMC should have written operating procedures and maintain written records of all its meetings.

5.5.3 When using electronic trial data handling and/or remote electronic trial data systems, the sponsor should:

(a) Ensure and document that the electronic data processing system(s) conforms to the sponsor's established requirements for completeness, accuracy, reliability, and consistent intended performance (i.e., validation).

(b) Maintain SOPs for using these systems.

(c) Ensure that the systems are designed to permit data changes in such a way that the data changes are documented and that there is no deletion of entered data (i.e., maintain an audit trail, data trail, edit trail).

(d) Maintain a security system that prevents unauthorized access to the data.

(e) Maintain a list of the individuals who are authorized to make data changes (see Sections 4.1.5 and 4.9.3).

(f) Maintain adequate backup of the data.

(g) Safeguard the blinding, if any (for example, maintain the blinding during data entry and processing).

5.5.4 If data are transformed during processing, it should always be possible to compare the original data and observations with the processed data.

5.5.5 The sponsor should use an unambiguous subject identification code (see Section 1.58) that allows identification of all the data reported for each subject.

5.5.6 The sponsor, or other owners of the data, should retain all of the sponsor specific essential documents pertaining to the trial. (See

Section 8. "Essential Documents for the Conduct of a Clinical Trial.")

5.5.7 The sponsor should retain all sponsor-specific essential documents in conformance with the applicable regulatory requirement(s) of the country(ies) where the product is approved, and/or where the sponsor intends to apply for approval(s).

5.5.8 If the sponsor discontinues the clinical development of an investigational product (i.e., for any or all indications, routes of administration, or dosage forms), the sponsor should maintain all sponsor-specific essential documents for at least 2 years after formal discontinuation or in conformance with the applicable regulatory requirement(s).

5.5.9 If the sponsor discontinues the clinical development of an investigational product, the sponsor should notify all the trial investigators/institutions and all the appropriate regulatory authorities.

5.5.10 Any transfer of ownership of the data should be reported to the appropriate authority(ies), as required by the applicable regulatory requirement(s).

5.5.11 The sponsor-specific essential documents should be retained until at least 2 years after the last approval of a marketing application in an ICH region and until there are no pending or contemplated marketing applications in an ICH region or at least 2 years have elapsed since the formal discontinuation of clinical development of the investigational product. These documents should be retained for a longer period, however, if required by the applicable regulatory requirement(s) or if needed by the sponsor.

5.5.12 The sponsor should inform the investigator(s)/institution(s) in writing of the need for record retention and should notify the investigator(s)/institution(s) in writing when the trial-related records are no longer needed (see Section 4.9.5).

5.6 Investigator Selection

5.6.1 The sponsor is responsible for selecting the investigator(s)/institution(s). Each investigator should be qualified by training and experience and should have adequate resources (see Sections 4.1 and 4.2) to properly conduct the trial for which the investigator is selected. If a coordinating committee and/or coordinating investigator(s) are to be utilized in multicenter trials, their organization and/or selection are the sponsor's responsibility.

5.6.2 Before entering an agreement with an investigator/institution to conduct a trial, the sponsor should provide the investigator(s)/institution(s) with the protocol and an up-to-date Investigator's Brochure, and should provide sufficient time for the investigator/institution to review the protocol and the information provided.

5.6.3 The sponsor should obtain the investigator's/institution's agreement:

(a) To conduct the trial in compliance with GCP, with the applicable regulatory requirement(s), and with the protocol agreed to by the sponsor and given approval/favorable opinion by the IRB/IEC;

(b) To comply with procedures for data recording/reporting: and

(c) To permit monitoring, auditing, and inspection (see Section 4.1.4).

(d) To retain the essential documents that should be in the investigator/institution files (see Section 8) until the sponsor informs the investigator/institution these documents are no longer needed (see Sections 4.9.4, 4.9.5, and 5.5.12). The sponsor and the investigator/institution should sign the protocol, or an alternative document, to confirm this agreement.

5.7 Allocation of Duties and Functions

Prior to initiating a trial, the sponsor should define, establish, and allocate all trial-related duties and functions.

5.8 Compensation to Subjects and Investigators

5.8.1 If required by the applicable regulatory requirement(s), the sponsor should provide insurance or should indemnify (legal and financial coverage) the investigator/the institution against claims arising from the trial, except for claims that arise from malpractice and/or negligence.

5.8.2 The sponsor's policies and procedures should address the costs of treatment of trial subjects in the event of trial-related injuries in accordance with the applicable regulatory requirement(s).

5.8.3 When trial subjects receive compensation, the method and manner of compensation should comply with applicable regulatory requirement(s).

5.9 Financing

The financial aspects of the trial should be documented in an agreement between the sponsor and the investigator/institution.

5.10 Notification/Submission to Regulatory Authority(ies)

Before initiating the clinical trial(s), the sponsor (or the sponsor and the investigator, if required by the applicable regulatory requirement(s)), should submit any required application(s) to the appropriate authority(ies) for review, acceptance, and/or permission (as required by the applicable regulatory requirement(s)) to begin the trial(s). Any notification/submission should be dated and contain sufficient information to identify the protocol.

5.11 Confirmation of Review by IRB/IEC

5.11.1 The sponsor should obtain from the investigator/institution:

(a) The name and address of the investigator's/institution's IRB/IEC.

(b) A statement obtained from the IRB/IEC that it is organized and operates according to GCP and the applicable laws and regulations.

(c) Documented IRB/IEC approval/favorable opinion and, if requested by the sponsor, a current copy of protocol, written informed consent form(s) and any other written information to be provided to subjects, subject recruiting procedures, and documents related to payments and compensation available to the subjects, and any other documents that the IRB/IEC may have requested.

5.11.2 If the IRB/IEC conditions its approval/favorable opinion upon change(s) in any aspect of the trial, such as modification(s) of the protocol, written informed consent form and any other written information to be provided to subjects, and/or other procedures, the sponsor should obtain from the investigator/institution a copy of the modification(s) made and the date approval/favorable opinion was given by the IRB/IEC.

5.11.3 The sponsor should obtain from the investigator/institution documentation and dates of any IRB/IEC reapprovals/reevaluations with favorable opinion, and of any withdrawals or suspensions of approval/favorable opinion.

5.12 Information on Investigational Product(s)

5.12.1 When planning trials, the sponsor should ensure that sufficient safety and efficacy data from nonclinical studies and/or clinical trials are available to support human exposure by the route, at the dosages, for the duration, and in the trial population to be studied.

5.12.2 The sponsor should update the Investigator's Brochure as significant new information becomes available. (See Section 7. "Investigator's Brochure.")

5.13 Manufacturing, Packaging, Labeling, and Coding Investigational Product(s)

5.13.1 The sponsor should ensure that the investigational product(s) (including active comparator(s) and placebo, if applicable) is characterized as appropriate to the stage of development of the product(s), is manufactured in accordance with any applicable GMP, and is coded and labeled in a manner that protects the blinding, if applicable. In addition, the labeling should comply with applicable regulatory requirement(s).

5.13.2 The sponsor should determine, for the investigational product(s), acceptable storage temperatures, storage conditions (for example, protection from light), storage times, reconstitution fluids and procedures, and devices for product infusion, if any. The sponsor should inform all involved parties (monitors, investigators, pharmacists, storage managers) of these determinations.

5.13.3 The investigational product(s) should be packaged to prevent contamination and unacceptable deterioration during transport and storage.

5.13.4 In blinded trials, the coding system for the investigational product(s) should include a mechanism that permits rapid identification of the product(s) in case of a medical emergency, but does not permit undetectable breaks of the blinding.

5.13.5 If significant formulation changes are made in the investigational or comparator product(s) during the course of clinical development, the results of any additional studies of the formulated product(s) (for example, stability, dissolution rate, bioavailability) needed to assess whether these changes would significantly alter the pharmacokinetic profile of the product should be available prior to the use of the new formulation in clinical trials.

5.14 Supplying and Handling Investigational Product(s)

5.14.1 The sponsor is responsible for supplying the investigator(s)/institution(s) with the investigational product(s).

5.14.2 The sponsor should not supply an investigator/institution with the investigational product(s) until the sponsor obtains all required documentation (for example, approval/favorable opinion from IRB/IEC and regulatory authority(ies)).

5.14.3 The sponsor should ensure that written procedures include instructions that the investigator/institution should follow for the handling and storage of investigational product(s) for the trial and documentation thereof. The procedures should address adequate and safe receipt, handling, storage, dispensing, retrieval of unused product from subjects, and return of unused investigational product(s) to the sponsor (or alternative disposition if authorized by the sponsor and in compliance with the applicable regulatory requirement(s)).

5.14.4 The sponsor should:

(a) Ensure timely delivery of investigational product(s) to the investigator(s).

(b) Maintain records that document shipment, receipt, disposition, return, and destruction of the investigational product(s). (See Section 8. "Essential Documents for the Conduct of a Clinical Trial.")

(c) Maintain a system for retrieving investigational products and documenting this retrieval (for example, for deficient product recall, reclaim after trial completion, expired product reclaim).

(d) Maintain a system for the disposition of unused investigational product(s) and for the documentation of this disposition.

5.14.5 The sponsor should:

(a) Take steps to ensure that the investigational product(s) are stable over the period of use.

(b) Maintain sufficient quantities of the investigational product(s) used in the trials to reconfirm specifications, should this become necessary, and maintain records of batch sample analyses and characteristics. To the extent stability permits, samples should be retained either until the analyses of the trial data are complete or as required by the applicable regulatory requirement(s), whichever represents the longer retention period.

5.15 Record Access

5.15.1 The sponsor should ensure that it is specified in the protocol or other written agreement that the investigator(s)/institution(s) provide direct access to source data/documents for trial-related monitoring, audits, IRB/IEC review, and regulatory inspection.

5.15.2 The sponsor should verify that each subject has consented, in writing, to direct access to his/her original medical records for trial-related monitoring, audit, IRB/IEC review, and regulatory inspection.

5.16 Safety Information

5.16.1 The sponsor is responsible for the ongoing safety evaluation of the investigational product(s).

5.16.2 The sponsor should promptly notify all concerned investigator(s)/institution(s) and the regulatory authority(ies) of findings that could affect adversely the safety of subjects, impact the conduct of the trial, or alter the IRB/IEC's approval/favorable opinion to continue the trial.

5.17 Adverse Drug Reaction Reporting

5.17.1 The sponsor should expedite the reporting to all concerned investigator(s)/institutions(s), to the IRB(s)/IEC(s), where required, and to the regulatory authority(ies) of all adverse drug reactions (ADRs) that are both serious and unexpected.

5.17.2 Such expedited reports should comply with the applicable regulatory requirement(s) and with the ICH Guidance for Clinical Safety Data Management: Definitions and Standards for Expedited Reporting.

5.17.3 The sponsor should submit to the regulatory authority(ies) all safety updates and periodic reports, as required by applicable regulatory requirement(s).

5.18 Monitoring

5.18.1 **Purpose**

The purposes of trial monitoring are to verify that:

(a) The rights and well-being of human subjects are protected.
(b) The reported trial data are accurate, complete, and verifiable from source documents.
(c) The conduct of the trial is in compliance with the currently approved protocol/amendment(s), with GCP, and with applicable regulatory requirement(s).

5.18.2 **Selection and Qualifications of Monitors**

(a) Monitors should be appointed by the sponsor.
(b) Monitors should be appropriately trained, and should have the scientific and/or clinical knowledge needed to monitor the trial adequately. A monitor's qualifications should be documented.
(c) Monitors should be thoroughly familiar with the investigational product(s), the protocol, written informed consent

form and any other written information to be provided to subjects, the sponsor's SOPs, GCP, and the applicable regulatory requirement(s).

5.18.3 **Extent and Nature of Monitoring**

The sponsor should ensure that the trials are adequately monitored. The sponsor should determine the appropriate extent and nature of monitoring. The determination of the extent and nature of monitoring should be based on considerations such as the objective, purpose, design, complexity, blinding, size, and endpoints of the trial. In general there is a need for on-site monitoring, before, during, and after the trial; however, in exceptional circumstances the sponsor may determine that central monitoring in conjunction with procedures such as investigators' training and meetings, and extensive written guidance can assure appropriate conduct of the trial in accordance with GCP. Statistically controlled sampling may be an acceptable method for selecting the data to be verified.

5.18.4 **Monitor's Responsibilities**

The monitor(s), in accordance with the sponsor's requirements, should ensure that the trial is conducted and documented properly by carrying out the following activities when relevant and necessary to the trial and the trial site:

(a) Acting as the main line of communication between the sponsor and the investigator.

(b) Verifying that the investigator has adequate qualifications and resources (see Sections 4.1, 4.2, and 5.6) and these remain adequate throughout the trial period, and that the staff and facilities, including laboratories and equipment, are adequate to safely and properly conduct the trial and these remain adequate throughout the trial period.

(c) Verifying, for the investigational product(s):

   (i) That storage times and conditions are acceptable, and that supplies are sufficient throughout the trial.

   (ii) That the investigational product(s) are supplied only to subjects who are eligible to receive it and at the protocol specified dose(s).

   (iii) That subjects are provided with necessary instruction on properly using, handling, storing, and returning the investigational product(s).

   (iv) That the receipt, use, and return of the investigational product(s) at the trial sites are controlled and documented adequately.

   (v) That the disposition of unused investigational product(s) at the trial sites complies with applicable regulatory requirement(s) and is in accordance with the sponsor's authorized procedures.

(d) Verifying that the investigator follows the approved protocol and all approved amendment(s), if any.
(e) Verifying that written informed consent was obtained before each subject's participation in the trial.
(f) Ensuring that the investigator receives the current Investigator's Brochure, all documents, and all trial supplies needed to conduct the trial properly and to comply with the applicable regulatory requirement(s).
(g) Ensuring that the investigator and the investigator's trial staff are adequately informed about the trial.
(h) Verifying that the investigator and the investigator's trial staff are performing the specified trial functions, in accordance with the protocol and any other written agreement between the sponsor and the investigator/institution, and have not delegated these functions to unauthorized individuals.
(i) Verifying that the investigator is enrolling only eligible subjects.
(j) Reporting the subject recruitment rate.
(k) Verifying that source data/documents and other trial records are accurate, complete, kept up-to-date, and maintained.
(l) Verifying that the investigator provides all the required reports, notifications, applications, and submissions, and that these documents are accurate, complete, timely, legible, dated, and identify the trial.
(m) Checking the accuracy and completeness of the CRF entries, source data/documents, and other trial-related records against each other. The monitor specifically should verify that:
    (i) The data required by the protocol are reported accurately on the CRFs and are consistent with the source data/documents.
    (ii) Any dose and/or therapy modifications are well documented for each of the trial subjects.
    (iii) Adverse events, concomitant medications, and intercurrent illnesses are reported in accordance with the protocol on the CRFs.
    (iv) Visits that the subjects fail to make, tests that are not conducted, and examinations that are not performed are clearly reported as such on the CRFs.
    (v) All withdrawals and dropouts of enrolled subjects from the trial are reported and explained on the CRFs.
(n) Informing the investigator of any CRF entry error, omission, or illegibility. The monitor should ensure that appropriate corrections, additions, or deletions are made, dated, explained (if necessary), and initialed by the investigator or by a member of the investigator's trial staff who is authorized

to initial CRF changes for the investigator. This authorization should be documented.

(o) Determining whether all adverse events (AEs) are appropriately reported within the time periods required by GCP, the ICH Guidance for Clinical Safety Data Management: Definitions and Standards for Expedited Reporting, the protocol, the IRB/IEC, the sponsor, and the applicable regulatory requirement(s).

(p) Determining whether the investigator is maintaining the essential documents. (See Section 8. "Essential Documents for the Conduct of a Clinical Trial.")

(q) Communicating deviations from the protocol, SOPs, GCP, and the applicable regulatory requirements to the investigator and taking appropriate action designed to prevent recurrence of the detected deviations.

5.18.5 **Monitoring Procedures**

The monitor(s) should follow the sponsor's established written SOPs as well as those procedures that are specified by the sponsor for monitoring a specific trial.

5.18.6 **Monitoring Report**

(a) The monitor should submit a written report to the sponsor after each trial-site visit or trial-related communication.

(b) Reports should include the date, site, name of the monitor, and name of the investigator or other individual(s) contacted.

(c) Reports should include a summary of what the monitor reviewed and the monitor's statements concerning the significant findings/facts, deviations and deficiencies, conclusions, actions taken or to be taken, and/or actions recommended to secure compliance.

(d) The review and follow-up of the monitoring report by the sponsor should be documented by the sponsor's designated representative.

5.19 Audit

If or when sponsors perform audits, as part of implementing quality assurance, they should consider:

5.19.1 **Purpose**

The purpose of a sponsor's audit, which is independent of and separate from routine monitoring or quality control functions, should be to evaluate trial conduct and compliance with the protocol, SOPs, GCP, and the applicable regulatory requirements.

5.19.2 **Selection and Qualification of Auditors**

(a) The sponsor should appoint individuals, who are independent of the clinical trial/data collection system(s), to conduct audits.

(b) The sponsor should ensure that the auditors are qualified by training and experience to conduct audits properly. An auditor's qualifications should be documented.

5.19.3 **Auditing Procedures**

(a) The sponsor should ensure that the auditing of clinical trials/systems is conducted in accordance with the sponsor's written procedures on what to audit, how to audit, the frequency of audits, and the form and content of audit reports.

(b) The sponsor's audit plan and procedures for a trial audit should be guided by the importance of the trial to submissions to regulatory authorities, the number of subjects in the trial, the type and complexity of the trial, the level of risks to the trial subjects, and any identified problem(s).

(c) The observations and findings of the auditor(s) should be documented.

(d) To preserve the independence and value of the audit function, the regulatory authority(ies) should not routinely request the audit reports. Regulatory authority(ies) may seek access to an audit report on a case-by case basis, when evidence of serious GCP noncompliance exists, or in the course of legal proceedings.

(e) Where required by applicable law or regulation, the sponsor should provide an audit certificate.

5.20 Noncompliance

5.20.1 Noncompliance with the protocol, SOPs, GCP, and/or applicable regulatory requirement(s) by an investigator/institution, or by member(s) of the sponsor's staff should lead to prompt action by the sponsor to secure compliance.

5.20.2 If the monitoring and/or auditing identifies serious and/or persistent noncompliance on the part of an investigator/institution, the sponsor should terminate the investigator's/institution's participation in the trial. When an investigator's/institution's participation is terminated because of noncompliance, the sponsor should notify promptly the regulatory authority(ies).

5.21 Premature Termination or Suspension of a Trial

If a trial is terminated prematurely or suspended, the sponsor should promptly inform the investigators/institutions, and the regulatory authority(ies) of the termination or suspension and the reason(s) for the termination or suspension. The IRB/IEC should also be informed promptly and provided the reason(s) for the termination or suspension by the sponsor or by the investigator/institution, as specified by the applicable regulatory requirement(s).

5.22 Clinical Trial/Study Reports

Whether the trial is completed or prematurely terminated, the sponsor should ensure that the clinical trial/study reports are prepared and provided to the regulatory agency(ies) as required by the applicable regulatory requirement(s). The sponsor should also ensure that the clinical trial/study reports in marketing applications meet the standards of the ICH Guidance for Structure and Content of Clinical Study Reports. (Note: The ICH

Guidance for Structure and Content of Clinical Study Reports specifies that abbreviated study reports may be acceptable in certain cases.)

5.23 Multicenter Trials

For multicenter trials, the sponsor should ensure that:

5.23.1 All investigators conduct the trial in strict compliance with the protocol agreed to by the sponsor and, if required, by the regulatory authority(ies), and given approval/favorable opinion by the IRB/IEC.

5.23.2 The CRFs are designed to capture the required data at all multicenter trial sites. For those investigators who are collecting additional data, supplemental CRFs should also be provided that are designed to capture the additional data.

5.23.3 The responsibilities of the coordinating investigator(s) and the other participating investigators are documented prior to the start of the trial.

5.23.4 All investigators are given instructions on following the protocol, on complying with a uniform set of standards for the assessment of clinical and laboratory findings, and on completing the CRFs.

5.23.5 Communication between investigators is facilitated.

6. CLINICAL TRIAL PROTOCOL AND PROTOCOL

The contents of a trial protocol should generally include the following topics. However, site specific information may be provided on separate protocol page(s), or addressed in a separate agreement, and some of the information listed below may be contained in other protocol referenced documents, such as an Investigator's Brochure.

6.1 General Information

6.1.1 Protocol title, protocol identifying number, and date. Any amendment(s) should also bear the amendment number(s) and date(s).

6.1.2 Name and address of the sponsor and monitor (if other than the sponsor).

6.1.3 Name and title of the person(s) authorized to sign the protocol and the protocol amendment(s) for the sponsor.

6.1.4 Name, title, address, and telephone number(s) of the sponsor's medical expert (or dentist when appropriate) for the trial.

6.1.5 Name and title of the investigator(s) who is (are) responsible for conducting the trial, and the address and telephone number(s) of the trial site(s).

6.1.6 Name, title, address, and telephone number(s) of the qualified physician (or dentist, if applicable) who is responsible for all trial-site related medical (or dental) decisions (if other than investigator).

6.1.7 Name(s) and address(es) of the clinical laboratory(ies) and other medical and/or technical department(s) and/or institutions involved in the trial.

6.2 Background Information

6.2.1 Name and description of the investigational product(s).

6.2.2 A summary of findings from nonclinical studies that potentially have clinical significance and from clinical trials that are relevant to the trial.

6.2.3 Summary of the known and potential risks and benefits, if any, to human subjects.

6.2.4 Description of and justification for the route of administration, dosage, dosage regimen, and treatment period(s).

6.2.5 A statement that the trial will be conducted in compliance with the protocol, GCP, and the applicable regulatory requirement(s).

6.2.6 Description of the population to be studied.

6.2.7 References to literature and data that are relevant to the trial, and that provide background for the trial.

6.3 Trial Objectives and Purpose

A detailed description of the objectives and the purpose of the trial.

6.4 Trial Design

The scientific integrity of the trial and the credibility of the data from the trial depend substantially on the trial design. A description of the trial design should include:

6.4.1 A specific statement of the primary endpoints and the secondary endpoints, if any, to be measured during the trial.

6.4.2 A description of the type/design of trial to be conducted (for example, double-blind, placebo-controlled, parallel design) and a schematic diagram of trial design, procedures, and stages.

6.4.3 A description of the measures taken to minimize/avoid bias, including (for example):

(a) Randomization.

(b) Blinding.

6.4.4 A description of the trial treatment(s) and the dosage and dosage regimen of the investigational product(s). Also include a description of the dosage form, packaging, and labeling of the investigational product(s).

6.4.5 The expected duration of subject participation, and a description of the sequence and duration of all trial periods, including follow-up, if any.

6.4.6 A description of the "stopping rules" or "discontinuation criteria" for individual subjects, parts of trial, and entire trial.

6.4.7 Accountability procedures for the investigational product(s), including the placebo(s) and comparator(s), if any.

6.4.8 Maintenance of trial treatment randomization codes and procedures for breaking codes.

6.4.9 The identification of any data to be recorded directly on the CRFs (i.e., no prior written or electronic record of data), and to be considered to be source data.

6.5 Selection and Withdrawal of Subjects

6.5.1 Subject inclusion criteria.

6.5.2 Subject exclusion criteria.

6.5.3 Subject withdrawal criteria (i.e., terminating investigational product treatment/trial treatment) and procedures specifying:

(a) When and how to withdraw subjects from the trial/investigational product treatment.

(b) The type and timing of the data to be collected for withdrawn subjects.

(c) Whether and how subjects are to be replaced.

(d) The follow-up for subjects withdrawn from investigational product treatment/trial treatment.

6.6 Treatment of Subjects

6.6.1 The treatment(s) to be administered, including the name(s) of all the product(s), the dose(s), the dosing schedule(s), the route/mode(s) of administration, and the treatment period(s), including the follow-up period(s) for subjects for each investigational product treatment/trial treatment group/arm of the trial.

6.6.2 Medication(s)/treatment(s) permitted (including rescue medication) and not permitted before and/or during the trial.

6.6.3 Procedures for monitoring subject compliance.

6.7 Assessment of Efficacy

6.7.1 Specification of the efficacy parameters.

6.7.2 Methods and timing for assessing, recording, and analyzing efficacy parameters.

6.8 Assessment of Safety

6.8.1 Specification of safety parameters.

6.8.2 The methods and timing for assessing, recording, and analyzing safety parameters.

6.8.3 Procedures for eliciting reports of and for recording and reporting adverse event and intercurrent illnesses.

6.8.4 The type and duration of the follow-up of subjects after adverse events.

6.9 Statistics

6.9.1 A description of the statistical methods to be employed, including timing of any planned interim analysis(ses).

6.9.2 The number of subjects planned to be enrolled. In multicenter trials, the number of enrolled subjects projected for each trial site should be specified. Reason for choice of sample size, including reflections on (or calculations of) the power of the trial and clinical justification.

6.9.3 The level of significance to be used.

6.9.4 Criteria for the termination of the trial.

6.9.5 Procedure for accounting for missing, unused, and spurious data.

6.9.6 Procedures for reporting any deviation(s) from the original statistical plan (any deviation(s) from the original statistical plan should be described and justified in the protocol and/or in the final report, as appropriate).

6.9.7 The selection of subjects to be included in the analyses (for example, all randomized subjects, all dosed subjects, all eligible subjects, evaluate-able subjects).

6.10 Direct Access to Source Data/Documents

The sponsor should ensure that it is specified in the protocol or other written agreement that the investigator(s)/institution(s) will permit trial-related monitoring, audits, IRB/IEC review, and regulatory inspection(s) by providing direct access to source data/documents.

6.11 Quality Control and Quality Assurance

6.12 Ethics

Description of ethical considerations relating to the trial.

6.13 Data Handling and Recordkeeping

6.14 Financing and Insurance

Financing and insurance if not addressed in a separate agreement.

6.15 Publication Policy

Publication policy, if not addressed in a separate agreement.

6.16 Supplements

(Note: Since the protocol and the clinical trial/study report are closely related, further relevant information can be found in the ICH Guidance for Structure and Content of Clinical Study Reports.)

7. INVESTIGATOR'S BROCHURE

7.1 Introduction

The Investigator's Brochure (IB) is a compilation of the clinical and nonclinical data on the investigational product(s) that are relevant to the study of the product(s) in human subjects. Its purpose is to provide the investigators and others involved in the trial with the information to facilitate their understanding of the rationale for, and their compliance with, many key features of the protocol, such as the dose, dose frequency/interval, methods of administration, and safety monitoring procedures. The IB also provides insight to support the clinical management of the study subjects during the course of the clinical trial. The information should be presented in a concise, simple, objective, balanced, and nonpromotional form that enables a clinician, or potential investigator, to understand it and make his/her own unbiased risk-benefit assessment of the appropriateness of the proposed trial. For this reason, a medically qualified person should generally participate in the editing of an IB, but the contents of the IB should be approved by the disciplines that generated the described data. This guidance delineates the minimum information that should be included in an IB and provides suggestions for its layout. It is expected that the type and extent of information available will vary with the stage of development of the investigational product. If the investigational product is marketed and its pharmacology is widely understood by medical practitioners, an extensive IB may not be necessary. Where permitted by regulatory authorities, a basic product information brochure, package leaflet, or labeling may be an appropriate alternative, provided that it includes current, comprehensive, and detailed information on all aspects of the investigational product that might be of importance to the investigator. If a marketed product is being studied for a new use (i.e., a new indication), an IB specific to that new use should be prepared. The IB should be reviewed at least annually and revised as necessary in compliance with a sponsor's written procedures. More frequent revision may be appropriate depending on the stage of development and the generation of relevant new information. However, in accordance with GCP, relevant new information may be so important that it should be communicated to the investigators, and possibly to the Institutional Review Boards (IRBs)/ Independent Ethics Committees (IECs) and/or regulatory authorities before it is included in a revised IB.

Generally, the sponsor is responsible for ensuring that an up-to-date IB is made available to the investigator(s) and the investigators are responsible for providing the up-to-date IB to the responsible IRBs/ IECs. In the case of an investigator-sponsored trial, the sponsor investigator should determine whether a brochure is available from the commercial manufacturer. If the investigational product is provided by the sponsor-investigator, then he or she should provide the necessary information to the trial personnel. In cases where preparation of a formal IB is impractical, the sponsor-investigator should provide, as a substitute, an expanded background information section in the trial protocol that contains the minimum current information described in this guidance.

7.2 General Considerations

The IB should include:

7.2.1 **Title Page**

This should provide the sponsor's name, the identity of each investigational product (i.e., research number, chemical or approved generic name, and trade name(s) where legally permissible and desired by the sponsor), and the release date. It is also suggested that an edition number, and a reference to the number and date of the edition it supersedes, be provided. An example is given in Appendix 1.

7.2.2 **Confidentiality Statement**

The sponsor may wish to include a statement instructing the investigator/ recipients to treat the IB as a confidential document for the sole information and use of the investigator's team and the IRB/IEC.

7.3 Contents of the Investigator's Brochure

The IB should contain the following sections, each with literature references where appropriate:

7.3.1 **Table of Contents**

An example of the Table of Contents is given in Appendix 2.

7.3.2 **Summary**

A brief summary (preferably not exceeding two pages) should be given, highlighting the significant physical, chemical, pharmaceutical, pharmacological, toxicological, pharmacokinetic, metabolic, and clinical information available that is relevant to the stage of clinical development of the investigational product.

7.3.3 **Introduction**

A brief introductory statement should be provided that contains the chemical name (and generic and trade name(s) when approved) of the investigational product(s), all active ingredients, the investigational product(s) pharmacological class and its expected position within this class (for example, advantages), the rationale for performing research with the investigational product(s), and the anticipated prophylactic, therapeutic, or diagnostic indication(s). Finally, the introductory statement should provide the general approach to be followed in evaluating the investigational product.

7.3.4 **Physical, Chemical, and Pharmaceutical Properties and Formulation**

A description should be provided of the investigational product substance(s) (including the chemical and/or structural formula (e)), and a brief summary should be given of the relevant physical, chemical, and pharmaceutical properties. To permit appropriate safety measures to be taken in the course of the trial, a description of the formulation(s) to be used, including excipients, should be provided and justified if clinically relevant. Instructions for the storage and handling of the dosage form(s) should also be given. Any structural similarities to other known compounds should be mentioned.

7.3.5 **Nonclinical Studies**

Introduction:

The results of all relevant nonclinical pharmacology, toxicology, pharmacokinetic, and investigational product metabolism studies should be provided in summary form. This summary should address the methodology used, the results, and a discussion of the relevance of the findings to the investigated therapeutic and the possible unfavorable and unintended effects in humans.

The information provided may include the following, as appropriate, if known/available:

Species tested;

Number and sex of animals in each group;

Unit dose (for example, milligram/kilogram (mg/kg));

Dose interval;

Route of administration;

Duration of dosing;

Information on systemic distribution;

Duration of post-exposure follow-up;

Results, including the following aspects:

- Nature and frequency of pharmacological or toxic effects;
- Severity or intensity of pharmacological or toxic effects;

- Time to onset of effects;
- Reversibility of effects;
- Duration of effects;
- Dose response.

Tabular format/listings should be used whenever possible to enhance the clarity of the presentation.

The following sections should discuss the most important findings from the studies, including the dose response of observed effects, the relevance to humans, and any aspects to be studied in humans. If applicable, the effective and nontoxic dose findings in the same animal species should be compared (i.e., the therapeutic index should be discussed). The relevance of this information to the proposed human dosing should be addressed. Whenever possible, comparisons should be made in terms of blood/tissue levels rather than on a mg/kg basis.

(a) Nonclinical Pharmacology

A summary of the pharmacological aspects of the investigational product and, where appropriate, its significant metabolites studied in animals should be included. Such a summary should incorporate studies that assess potential therapeutic activity (for example, efficacy models, receptor binding, and specificity) as well as those that assess safety (for example, special studies to assess pharmacological actions other than the intended therapeutic effect(s)).

(b) Pharmacokinetics and Product Metabolism in Animals

A summary of the pharmacokinetics and biological transformation and disposition of the investigational product in all species studied should be given. The discussion of the findings should address the absorption and the local and systemic bioavailability of the investigational product and its metabolites, and their relationship to the pharmacological and toxicological findings in animal species.

(c) Toxicology

A summary of the toxicological effects found in relevant studies conducted in different animal species should be described under the following headings where appropriate:

Single dose;
Repeated dose;
Carcinogenicity;
Special studies (for example, irritancy and sensitization);
Reproductive toxicity;
Genotoxicity (mutagenicity).

7.3.6 **Effects in Humans**

Introduction:

A thorough discussion of the known effects of the investigational product(s) in humans should be provided, including information on pharmacokinetics, metabolism, pharmacodynamics, dose response, safety, efficacy, and other pharmacological activities.

Where possible, a summary of each completed clinical trial should be provided. Information should also be provided regarding results from any use of the investigational product(s) other than in clinical trials, such as from experience during marketing.

(a) Pharmacokinetics and Product Metabolism in Humans

A summary of information on the pharmacokinetics of the investigational product(s) should be presented, including the following, if available:

Pharmacokinetics (including metabolism, as appropriate, and absorption, plasma protein binding, distribution, and elimination).

Bioavailability of the investigational product (absolute, where possible, and/or relative) using a reference dosage form.

Population subgroups (for example, gender, age, and impaired organ function).

Interactions (for example, product-product interactions and effects of food).

Other pharmacokinetic data (for example, results of population studies performed within clinical trial(s)).

(b) Safety and Efficacy

A summary of information should be provided about the investigational product's/products' (including metabolites, where appropriate) safety, pharmacodynamics, efficacy, and dose response that were obtained from preceding trials in humans (healthy volunteers and/or patients). The implications of this information should be discussed. In cases where a number of clinical trials have been completed, the use of summaries of safety and efficacy across multiple trials by indications in subgroups may provide a clear presentation of the data. Tabular summaries of adverse drug reactions for all the clinical trials (including those for all the studied indications) would be useful. Important differences in adverse drug reaction patterns/incidences across indications or subgroups should be discussed.

The IB should provide a description of the possible risks and adverse drug reactions to be anticipated on the basis of prior experiences with the product under investigation and with related products. A description should also be provided of the precautions or special monitoring to be done as part of the investigational use of the product(s).

(c) Marketing Experience

The IB should identify countries where the investigational product has been marketed or approved. Any significant information arising from the marketed use should be summarized (for example, formulations, dosages, routes of administration, and adverse product reactions). The IB should also identify all the countries where the investigational product did not receive approval/registration for marketing or was withdrawn from marketing/registration.

7.3.7 **Summary of Data and Guidance for the Investigator**

This section should provide an overall discussion of the nonclinical and clinical data, and should summarize the information from various sources on different aspects of the investigational product(s), wherever possible. In this way, the investigator can be provided with the most informative interpretation of the available data and with an assessment of the implications of the information for future clinical trials. Where appropriate, the published reports on related products should be discussed. This could help the investigator to anticipate adverse drug reactions or other problems in clinical trials. The overall aim of this section is to provide the investigator with a clear understanding of the possible risks and adverse reactions, and of the specific tests, observations, and precautions that may be needed for a clinical trial. This understanding should be based on the available physical, chemical, pharmaceutical, pharmacological, toxicological, and clinical information on the investigational product(s). Guidance should also be provided to the clinical investigator on the recognition and treatment of possible overdose and adverse drug reactions that is based on previous human experience and on the pharmacology of the investigational product.

7.4 Appendix 1

TITLE PAGE OF INVESTIGATOR'S BROCHURE

7.5 Appendix 2

TABLE OF CONTENTS OF INVESTIGATOR'S BROCHURE

Confidentiality Statement (optional)
Signature Page (optional)
1 Table of Contents
2 Summary
3 Introduction
4 Physical, Chemical, and Pharmaceutical Properties and Formulation
5 Nonclinical Studies
    5.1 Nonclinical Pharmacology
    5.2 Pharmacokinetics and Product Metabolism in Animals
    5.3 Toxicology
6 Effects in Humans
    6.1 Pharmacokinetics and Product Metabolism in Humans
    6.2 Safety and Efficacy
    6.3 Marketing Experience
7 Summary of Data and Guidance for the Investigator
    NB: References on
    1. Publications
    2. Reports

These references should be found at the end of each chapter

Appendices (if any)

8. ESSENTIAL DOCUMENTS FOR THE CONDUCT OF A CLINICAL TRIAL

8.1 Introduction

8.2 Before the Clinical Phase of the Trial Commences

8.3 During the Clinical Conduct of the Trial

8.4 After Completion or Termination of the Trial

Web Links to India's Regulatory Acts and Guidelines

Sr. No.	Document	Link
1	CDSCO's Good Clinical Practice (GCP) guidelines, 2001	http://www.cdsco.nic.in/html/GCP1.html
2	Draft of Medical Devices Rules, 2016	http://cdsco.nic.in/writereaddata/Draft_Medical%20 Devices%20Rules%202016.pdf
3	Drugs and Cosmetics Act 1940 and Drugs and Cosmetics Rules 1945, as amended in 2005	http://www.cdsco.nic.in/writereaddata/Drugs&Cosmetic Act.pdf
4	Drugs and Cosmetics Act 1940 and Drugs and Cosmetics Rules 1945, as amended in 2013	http://www.mohfw.nic.in/WriteReadData/l892s /43503435431421382269.pdf
5	ICMR's Ethical Guidelines for Biomedical Research on Human Subjects, 2006	http://www.icmr.nic.in/ethical_guidelines.pdf
6	National Guidelines for Stem Cell Research, 2013	http://icmr.nic.in/guidelines/NGSCR%202013.pdf
7	Schedule Y	http://cdsco.nic.in/html/D&C_Rules_Schedule_Y.pdf

THE NUREMBERG CODE

1. The voluntary consent of the human subject is absolutely essential.

This means that the person involved should have legal capacity to give consent; should be so situated as to be able to exercise free power of choice, without the intervention of any element of force, fraud, deceit, duress, overreaching, or other ulterior form of constraint or coercion; and should have sufficient knowledge and comprehension of the elements of the subject matter involved, as to enable him to make an understanding and enlightened decision. This latter element requires that, before the acceptance of an affirmative decision by the experimental subject, there should be made known to him the nature, duration, and purpose of the experiment; the method and means by which it is to be conducted; all inconveniences and hazards reasonably to be expected; and the effects upon his health or person, which may possibly come from his participation in the experiment.

The duty and responsibility for ascertaining the quality of the consent rests upon each individual who initiates, directs or engages in the experiment. It is a personal duty and responsibility which may not be delegated to another with impunity.

2. The experiment should be such as to yield fruitful results for the good of society, unprocurable by other methods or means of study, and not random and unnecessary in nature.
3. The experiment should be so designed and based on the results of animal experimentation and a knowledge of the natural history of the disease or other problem under study, that the anticipated results will justify the performance of the experiment.
4. The experiment should be so conducted as to avoid all unnecessary physical and mental suffering and injury.
5. No experiment should be conducted, where there is an a priori reason to believe that death or disabling injury will occur; except, perhaps, in those experiments where the experimental physicians also serve as subjects.
6. The degree of risk to be taken should never exceed that determined by the humanitarian importance of the problem to be solved by the experiment.
7. Proper preparations should be made and adequate facilities provided to protect the experimental subject against even remote possibilities of injury, disability, or death.
8. The experiment should be conducted only by scientifically qualified persons. The highest degree of skill and care should be required through all stages of the experiment of those who conduct or engage in the experiment.
9. During the course of the experiment, the human subject should be at liberty to bring the experiment to an end, if he has reached the physical or mental state, where continuation of the experiment seemed to him to be impossible.
10. During the course of the experiment, the scientist in charge must be prepared to terminate the experiment at any stage, if he has probable cause to

believe, in the exercise of the good faith, superior skill and careful judgement required of him, that a continuation of the experiment is likely to result in injury, disability, or death to the experimental subject.

["Trials of War Criminals before the Nuremberg Military Tribunals under Control Council Law No. 10," Vol. 2, pp. 181–182. Washington, D.C.: U.S. Government Printing Office, 1949.]

SAMPLE FDA FORM 483—LIST OF OBSERVATIONS

DEPARTMENT OF HEALTH AND HUMAN SERVICES
FOOD AND DRUG ADMINISTRATION

DISTRICT OFFICE ADDRESS AND PHONE NUMBER	DATE(S) OF INSPECTION
158-15 Liberty Avenue Jamaica, New York, 11433 (718)340-7000; Fax (718)662-5661 Industry Information: www.fda.gov/oc/industry	05/18/2016-06/10/2016 FEI NUMBER 3005287250

NAME AND TITLE OF INDIVIDUAL TO WHOM REPORT IS ISSUED

TO: Hymie Aruch, RPh, Pharmacy Manager

FIRM NAME	STREET ADDRESS
Region Care, Inc. (Northwell Health)	200 Community Drive
CITY, STATE AND ZIP CODE	TYPE OF ESTABLISHMENT INSPECTED
Great Neck, New York, 11021	Producer of Sterile Products

THIS DOCUMENT LISTS OBSERVATIONS MADE BY THE FDA REPRESENTATIVE(S) DURING THE INSPECTION OF YOUR FACILITY. THEY ARE INSPECTIONAL OBSERVATIONS, AND DO NOT REPRESENT A FINAL AGENCY DETERMINATION REGARDING YOUR COMPLIANCE. IF YOU HAVE AN OBJECTION REGARDING AN OBSERVATION, OR HAVE IMPLEMENTED, OR PLAN TO IMPLEMENT CORRECTIVE ACTION IN RESPONSE TO AN OBSERVATION, YOU MAY DISCUSS THE OBJECTION OR ACTION WITH THE FDA REPRESENTATIVE(S) DURING THE INSPECTION OR SUBMIT THIS INFORMATION TO FDA AT THE ADDRESS ABOVE. IF YOU HAVE ANY QUESTIONS, PLEASE CONTACT FDA AT THE PHONE NUMBER AND ADDRESS ABOVE.

DURING AN INSPECTION OF YOUR FIRM (I) (WE) OBSERVED:

1. Procedures designed to prevent microbiological contamination of drug products purporting to be sterile do not include adequate validation of the sterilization process.

Specifically,

a. Smoke studies, performed by an outside vendor (b) (4), are not conducted under dynamic conditions that simulate routine aseptic operations. Consequently, there is no assurance that uninterrupted unidirectional laminar airflow is maintained during aseptic operations.

b. The HVAC returns in Cleanroom (b) (4) were observed to be obstructed by wire racks containing drug components and other supplies, as well as by garbage cans.

2. Drugs products purporting to be sterile and pyrogen-free are not laboratory tested to determine conformance with such requirements.

Specifically,

Sterile drug products prepared from non-sterile starting materials (e.g. morphine, hydromorphone, bupivacaine, and baclofen) are not tested for sterility and bacterial endotoxins (pyrogens) prior to release.

Add Continuation Page

SEE REVERSE OF THIS PAGE	EMPLOYEE(S) SIGNATURE	EMPLOYEE(S) NAME AND TITLE (*Print or Type*)	DATE ISSUED
	[signature]	Robert C. Steyert, Investigator	06/10/2016

FORM FDA 483 (9/08) PREVIOUS EDITION OBSOLETE INSPECTIONAL OBSERVATIONS Page 1 of 4

DEPARTMENT OF HEALTH AND HUMAN SERVICES
FOOD AND DRUG ADMINISTRATION

DISTRICT OFFICE ADDRESS AND PHONE NUMBER	DATE(S) OF INSPECTION
158-15 Liberty Avenue Jamaica, New York, 11433 (718)340-7000; Fax (718)662-5661 Industry Information: www.fda.gov/oc/industry	05/18/2016-06/10/2016 FEI NUMBER 3005287250

NAME AND TITLE OF INDIVIDUAL TO WHOM REPORT IS ISSUED

TO: Hymie Aruch, RPh, Pharmacy Manager

FIRM NAME	STREET ADDRESS
Region Care, Inc. (Northwell Health)	200 Community Drive
CITY, STATE AND ZIP CODE	TYPE OF ESTABLISHMENT INSPECTED
Great Neck, New York, 11021	Producer of Sterile Products

3. Aseptic processing areas are deficient regarding the system for monitoring environmental conditions.

Specifically,

a. Work surfaces, inside the ISO-5 workstations, are not sampled at least daily during periods of production. For example, review of the "(b) (4) Compounding Activity (b) (4) Log" found there was no such testing on the following days: 05/05/2016, 05/09/2016, 05/10/2016, 05/12/2016, 05/16/2016, and 05/19/2016. However, review of the prescription log indicated sterile drug products were prepared on each of these days.

b. Viable air sampling is not performed at least once daily on days of aseptic operations.

c. Magnehelic gauges (manometers) used for monitoring pressure differentials between classified rooms, including the ISO-7 classified cleanroom (which houses ISO-5 classified workstations) and the ISO-7 classified anteroom are not calibrated.

4. Laboratory controls do not include a determination of conformance to appropriate specifications for drug products.

Specifically,

There are no written procedures requiring the performance of visual checks of all sterile drug products for clarity, discoloration or particulates. Furthermore, there is no documentation to support this practice is actually being performed.

Add Continuation Page

SEE REVERSE OF THIS PAGE	EMPLOYEE(S) SIGNATURE	EMPLOYEE(S) NAME AND TITLE *(Print or Type)*	DATE ISSUED
		Robert C. Steyert, Investigator	06/10/2016

FORM FDA 483 (9/08) PREVIOUS EDITION OBSOLETE **INSPECTIONAL OBSERVATIONS** Page 2 of 4

DEPARTMENT OF HEALTH AND HUMAN SERVICES
FOOD AND DRUG ADMINISTRATION

DISTRICT OFFICE ADDRESS AND PHONE NUMBER	DATE(S) OF INSPECTION
158-15 Liberty Avenue Jamaica, New York, 11433 (718)340-7000; Fax (718)662-5661 Industry Information: www.fda.gov/oc/industry	05/18/2016-06/10/2016 FEI NUMBER 3005287250

NAME AND TITLE OF INDIVIDUAL TO WHOM REPORT IS ISSUED

TO: Hymie Aruch, RPh, Pharmacy Manager

FIRM NAME	STREET ADDRESS
Region Care, Inc. (Northwell Health)	200 Community Drive
CITY, STATE AND ZIP CODE	TYPE OF ESTABLISHMENT INSPECTED
Great Neck, New York, 11021	Producer of Sterile Products

5. Routine calibration of equipment is not performed according to a written program designed to assure proper performance.

Specifically,

(b) (4) , used to monitor temperature in the ISO-7 classified ante-room and the walk-in refrigerator (used to store drug products) are not calibrated. A sticker on each of the (b) (4) indicates they were last calibrated "03/2014."

6. Clothing of personnel engaged in compounding of sterile drug products is not appropriate for the duties they perform.

Specifically,

Sterile gowns worn by operators preparing sterile drug products are reused throughout the day (i.e. they are not immediately discarded when doffed).

7. There is a failure to thoroughly review any unexplained discrepancy and the failure of a batch or any of its components to meet any of its specifications whether or not the batch has been already distributed.

Specifically,

On at least two occasions, environmental monitoring excursions involving the detection of "single colony" growth from (b) (4)surface and (b) (4) floor samples, respectively were documented and investigated. However, these investigations were inadequate in that:

Add Continuation Page

SEE REVERSE OF THIS PAGE	EMPLOYEE(S) SIGNATURE	EMPLOYEE(S) NAME AND TITLE (*Print or Type*)	DATE ISSUED
		Robert C. Steyert, Investigator	06/10/2016

FORM FDA 483 (9/08) PREVIOUS EDITION OBSOLETE INSPECTIONAL OBSERVATIONS Page 3 of 4

DEPARTMENT OF HEALTH AND HUMAN SERVICES
FOOD AND DRUG ADMINISTRATION

DISTRICT OFFICE ADDRESS AND PHONE NUMBER	DATE(S) OF INSPECTION
158-15 Liberty Avenue Jamaica, New York, 11433 (718)340-7000; Fax (718)662-5661 Industry Information: www.fda.gov/oc/industry	05/18/2016-06/10/2016 FEI NUMBER 3005287250

NAME AND TITLE OF INDIVIDUAL TO WHOM REPORT IS ISSUED

TO: Hymie Aruch, RPh, Pharmacy Manager

FIRM NAME	STREET ADDRESS
Region Care, Inc. (Northwell Health)	200 Community Drive
CITY, STATE AND ZIP CODE	TYPE OF ESTABLISHMENT INSPECTED
Great Neck, New York, 11021	Producer of Sterile Products

- Remedial cleaning and retesting is not documented as being performed.
- They did not include an assessment to determine what compounding activities were occurring at the time and location of the excursion to identify any potentially adverse impact on drug product quality.
-The "Actions Taken" section of the Incident Reports state "(b) (4) " However, there is no documentation to support this was actually performed.

8. Aseptic processing areas are deficient regarding the system for cleaning and disinfecting the room and equipment to produce aseptic conditions.

Specifically,

a. The (b) (4) divider separating the LAFW (laminar airflow workstation) and (b) (4) is cleaned only (b) (4) .

b. An operator was observed dry mopping the floor of the ISO-7 classified ante-room, then enter the more critical ISO-7 classified clean room and continue mopping the floor without changing the mop head; instead of working from the greater to lesser critical area.

c. Cleaning procedures are deficient in that the contact time for disinfectants such as (b) (4) are not defined in written procedures or documented in cleaning records.

Add Continuation Page

SEE REVERSE OF THIS PAGE	EMPLOYEE(S) SIGNATURE	EMPLOYEE(S) NAME AND TITLE *(Print or Type)*	DATE ISSUED
	[signature]	Robert C. Steyert, Investigator	06/10/2016

FORM FDA 483 (9/08) PREVIOUS EDITION OBSOLETE INSPECTIONAL OBSERVATIONS Page 4 of 4

Summary of Federal Regulations Governing Human Research

Regulation	Description
21 CFR Part 50	This section of the CFR applies to all clinical investigations regulated by the FDA and regards the protection of the research subject. An investigator may not involve a human being in a research study without first obtaining a legally effective consent. Section 50.20 outlines the process for obtaining the consent and requires that the informed consent not contain language through which a subject is made to waive or appears to waive any of their legal rights. Not only is the investigator responsible for obtaining a proper and legal consent it is critical that appropriate portions of the consent be reflected in the CTA.
21 CFR Part 54	The FDA requires that each investigator conducting an FDA regulated study disclose any financial arrangements with the sponsor and the nature of those arrangements or certify they have no financial conflicts. It is the responsibility of the sponsor to collect a financial disclosure form from each investigator. The CTA should contain language that defines the process for completing this form, when it is due, and how the sponsor will collect the information.
21 CFR Part 312.62	This regulation, as well as Good Clinical Practices (GCPs), requires that the investigator maintain accurate records of the disposition of the drug and to retain these records for a period of two years after the FDA approves marketing of the drug or the study is terminated. The sponsor should define in the CTA how the records are to be stored and after approval of the drug or termination of the study their expectations for retaining records and how and when records may be destroyed.
21 CFR Part 56	The responsibilities of the Institutional Review Board (IRB), functions and operations, the board make-up, and criteria for protocol approval are addressed in this section of the CFR. The IRB, like the investigator, is tasked with the protection of the subject and ensuring that risks to the subject are minimized as much as possible. The IRB requirements and the party responsible for the requirement (for example, obtaining approval of the protocol and/or consent, reporting adverse events, and submitting continuing reviews) must be included in the CTA.
The Health Insurance Portability and Accountability Act (HIPAA)	This rule protects the privacy of individually identifiable health information. It also requires that the patient be informed of uses and disclosures of their medical information for research purposes, and their right to access information about them (45 CFR Parts 164.501, 164.508, and 164.512(i)). The CTA must contain language that the sponsor, the organization, and the investigative staff will comply with HIPAA.
21 CFR Part 312.68	The investigator upon the request of the FDA must allow authorized access to copy and verify all records and reports of the investigator. In the CTA the sponsor outlines their expectations of what the investigator should do when the FDA notifies them of an upcoming inspection. The CTA will cover when and how to notify the sponsor and how the sponsor will assist the site in preparing for an FDA inspection.
21 CFR 11	This regulation sets out the requirements for the use of electronic records to collect subject data and electronic signatures. The CTA should define how the data will be collected, verification of the investigator's signature, and the training on the electronic data capture system that will be provided to the investigator and study staff.
FDA Form 1572: Statement of the Investigator	Describes the investigators duties (U.S. Dept. of HHS) and agreement to comply with all requirements regarding the obligations of clinical investigators and all other pertinent requirements in 21 CFR 312. The investigator cannot begin a study prior to signing this statement. The CTA must correspond with the responsibilities and duties defined on this form.

18 USC Sec. 1001	This United States Code (USC) defines crimes and criminal procedures for any investigator or a member of the study team who knowingly commits fraud or provides false statements during the conduct of the research study. This should be spelled out clearly in the CTA.
The Medical Kickback Law – 42 U.S.C. §1320a-7b(b)	This statute prohibits companies paying or providing incentives to physicians to induce referrals. The CTA should include language related to physician incentives and state that the investigator cannot use incentives to induce physicians to refer patients to the study.
No Charge – 21 CFR 312 & 316	Study subjects cannot be charged for the investigational drug or procedures related to the study. This information should be included not only in the CTA but also the informed consent form.
Fair Market Value – 42 CFR 411 and 412, Affordable Care Act, 31 U.S.C. §3729-3733	Compensation of the investigator and the site must be reasonable and based on work. When negotiating the CTA and budget, the site/investigator and the sponsor must take into account what is reasonable compensation based on comparable services in the geographical area where the study is being conducted.
Debarment – 21 U.S.C. 335(a)m (b)(1) and (b)(2)	Investigators who have been found to have engaged in 'deliberate or repeated' violations of IND requirements may be debarred by the FDA resulting in the FDA denying them access to investigational drugs. The sponsor, the organization, and the investigator must certify that no individual involved in any aspect of the study has been debarred. The contract needs to contain language that all parties will exercise appropriate actions to ensure that a debarred investigator is not involved in the trial in any capacity.
Physician Open Payment Act – Affordable Care Act	Beginning in January 2013, under the Affordable Care Act, pharmaceutical companies must report to the Centers for Medicare & Medical Services payments of more than $10 made to physicians. This ACT is an effort to bring transparency to company-physician relationships. Under this Act the company must report aggregate information pertaining to individual healthcare professionals. This should also be included in CTAs signed after January 2013.

JPP/2014

THE BELMONT REPORT

Office of the Secretary
Ethical Principles and Guidelines for the Protection of Human Subjects of Research The National Commission for the Protection of Human Subjects of Biomedical and Behavioral Research

April 18, 1979

AGENCY: Department of Health, Education, and Welfare.

ACTION: Notice of Report for Public Comment.

SUMMARY: On July 12, 1974, the National Research Act (Pub. L. 93-348) was signed into law, there-by creating the National Commission for the Protection of Human Subjects of Biomedical and Behavioral Research. One of the charges to the Commission was to identify the basic ethical principles that should underlie the conduct of biomedical and behavioral research involving human subjects and to develop guidelines which should be followed to assure that such research is conducted in accordance with those principles. In carrying out the above, the Commission was directed to consider: (i) the boundaries between biomedical and behavioral research and the accepted and routine practice of medicine, (ii) the role of assessment of risk–benefit criteria in the determination of the appropriateness of research involving human subjects, (iii) appropriate guidelines for the selection of human subjects for participation in such research and (iv) the nature and definition of informed consent in various research settings.

The Belmont Report attempts to summarize the basic ethical principles identified by the Commission in the course of its deliberations. It is the outgrowth of an intensive four-day period of discussions that were held in February 1976 at the Smithsonian Institution's Belmont Conference Center supplemented by the monthly deliberations of the Commission that were held over a period of nearly four years. It is a statement of basic ethical principles and guidelines that should assist in resolving the ethical problems that surround the conduct of research with human subjects. By publishing the Report in the Federal Register, and providing reprints upon request, the Secretary intends that it may be made readily available to scientists, members of Institutional Review Boards, and Federal employees. The two-volume Appendix, containing the lengthy reports of experts and specialists who assisted the Commission in fulfilling this part of its charge, is available as DHEW Publication No. (OS) 78-0013 and No. (OS) 78-0014, for sale by the Superintendent of Documents, U.S. Government Printing Office, Washington, D.C. 20402.

Unlike most other reports of the Commission, the Belmont Report does not make specific recommendations for administrative action by the Secretary of Health, Education, and Welfare. Rather, the Commission recommended that the Belmont Report be adopted in its entirety, as a statement of the Department's policy. The Department requests public comment on this recommendation.

NATIONAL COMMISSION FOR THE PROTECTION OF HUMAN SUBJECTS OF BIOMEDICAL AND BEHAVIORAL RESEARCH

Members of the Commission

- Kenneth John Ryan, M.D., Chairman, Chief of Staff, Boston Hospital for Women.
- Joseph V. Brady, Ph.D., Professor of Behavioral Biology, Johns Hopkins University.
- Robert E. Cooke, M.D., President, Medical College of Pennsylvania.
- Dorothy I. Height, President, National Council of Negro Women, Inc.
- Albert R. Jonsen, Ph.D., Associate Professor of Bioethics, University of California at San Francisco.
- Patricia King, J.D., Associate Professor of Law, Georgetown University Law Center.
- Karen Lebacqz, Ph.D., Associate Professor of Christian Ethics, Pacific School of Religion.
- **** David W. Louisell, J.D., Professor of Law, University of California at Berkeley.*
- Donald W. Seldin, M.D., Professor and Chairman, Department of Internal Medicine, University of Texas at Dallas.
- ****Eliot Stellar, Ph.D., Provost of the University and Professor of Physiological Psychology, University of Pennsylvania.*
- **** Robert H. Turtle, LL.B., Attorney, VomBaur, Coburn, Simmons & Turtle, Washington, D.C.*
- ****Deceased.*

TABLE OF CONTENTS

ETHICAL PRINCIPLES AND GUIDELINES FOR RESEARCH INVOLVING HUMAN SUBJECTS

Scientific research has produced substantial social benefits. It has also posed some troubling ethical questions. Public attention was drawn to these questions by reported abuses of human subjects in biomedical experiments, especially during the Second

World War. During the Nuremberg War Crime Trials, the Nuremberg code was drafted as a set of standards for judging physicians and scientists who had conducted biomedical experiments on concentration camp prisoners. This code became the prototype of many later codes [1] intended to assure that research involving human subjects would be carried out in an ethical manner.

The codes consist of rules, some general, others specific that guide the investigators or the reviewers of research in their work. Such rules often are inadequate to cover complex situations; at times they come into conflict, and they are frequently difficult to interpret or apply. Broader ethical principles will provide a basis on which specific rules may be formulated, criticized and interpreted.

Three principles, or general prescriptive judgments, that are relevant to research involving human subjects are identified in this statement. Other principles may also be relevant. These three are comprehensive, however, and are stated at a level of generalization that should assist scientists, subjects, reviewers and interested citizens to understand the ethical issues inherent in research involving human subjects. These principles cannot always be applied so as to resolve beyond dispute particular ethical problems. The objective is to provide an analytical framework that will guide the resolution of ethical problems arising from research involving human subjects.

This statement consists of a distinction between research and practice, a discussion of the three basic ethical principles, and remarks about the application of these principles.

PART A: BOUNDARIES BETWEEN PRACTICE AND RESEARCH

It is important to distinguish between biomedical and behavioral research, on the one hand, and the practice of accepted therapy on the other, in order to know what activities ought to undergo review for the protection of human subjects of research. The distinction between research and practice is blurred partly because both often occur together (as in research designed to evaluate a therapy) and partly because notable departures from standard practice are often called "experimental" when the terms "experimental" and "research" are not carefully defined.

For the most part, the term "practice" refers to interventions that are designed solely to enhance the well-being of an individual patient or client and that have a reasonable expectation of success. The purpose of medical or behavioral practice is to provide diagnosis, preventive treatment or therapy to particular individuals [2]. By contrast, the term "research" designates an activity designed to test an hypothesis, permit conclusions to be drawn, and thereby to develop or contribute to generalizable knowledge (expressed, for example, in theories, principles, and statements of relationships). Research is usually described in a formal protocol that sets forth an objective and a set of procedures designed to reach that objective.

When a clinician departs in a significant way from standard or accepted practice, the innovation does not, in and of itself, constitute research. The fact that a procedure is "experimental," in the sense of new, untested or different, does not automatically place it in the category of research. Radically new procedures of this description should, however, be made the object of formal research at an early stage in order to determine whether they are safe and effective. Thus, it is the responsibility of medical practice committees, for example, to insist that a major innovation be incorporated into a formal research project [3].

Research and practice may be carried on together when research is designed to evaluate the safety and efficacy of a therapy. This need not cause any confusion regarding whether or not the activity requires review; the general rule is that if there is any element of research in an activity, that activity should undergo review for the protection of human subjects.

PART B: BASIC ETHICAL PRINCIPLES

The expression "basic ethical principles" refers to those general judgments that serve as a basic justification for the many particular ethical prescriptions and evaluations of human actions.

Three basic principles, among those generally accepted in our cultural tradition, are particularly relevant to the ethics of research involving human subjects: the principles of respect of persons, beneficence and justice.

1. **Respect for Persons**—Respect for persons incorporates at least two ethical convictions: first, that individuals should be treated as autonomous agents, and second, that persons with diminished autonomy are entitled to protection. The principle of respect for persons thus divides into two separate moral requirements: the requirement to acknowledge autonomy and the requirement to protect those with diminished autonomy.

   An autonomous person is an individual capable of deliberation about personal goals and of acting under the direction of such deliberation. To respect autonomy is to give weight to autonomous persons' considered opinions and choices while refraining from obstructing their actions unless they are clearly detrimental to others. To show lack of respect for an autonomous agent is to repudiate that person's considered judgments, to deny an individual the freedom to act on those considered judgments, or to withhold information necessary to make a considered judgment, when there are no compelling reasons to do so.

   However, not every human being is capable of self-determination. The capacity for self-determination matures during an individual's life, and some individuals lose this capacity wholly or in part because of illness, mental disability, or circumstances that severely restrict liberty.

   Respect for the immature and the incapacitated may require protecting them as they mature or while they are incapacitated.

   Some persons are in need of extensive protection, even to the point of excluding them from activities which may harm them; other persons require little protection beyond making sure they undertake activities freely and with awareness of possible adverse consequence. The extent of protection afforded should depend upon the risk of harm and the likelihood of benefit.

The judgment that any individual lacks autonomy should be periodically reevaluated and will vary in different situations.

In most cases of research involving human subjects, respect for persons demands that subjects enter into the research voluntarily and with adequate information. In some situations, however, application of the principle is not obvious. The involvement of prisoners as subjects of research provides an instructive example. On the one hand, it would seem that the principle of respect for persons requires that prisoners not be deprived of the opportunity to volunteer for research. On the other hand, under prison conditions they may be subtly coerced or unduly influenced to engage in research activities for which they would not otherwise volunteer. Respect for persons would then dictate that prisoners be protected. Whether to allow prisoners to "volunteer" or to "protect" them presents a dilemma. Respecting persons, in most hard cases, is often a matter of balancing competing claims urged by the principle of respect itself.

2. **Beneficence**—Persons are treated in an ethical manner not only by respecting their decisions and protecting them from harm, but also by making efforts to secure their well-being. Such treatment falls under the principle of beneficence. The term "beneficence" is often understood to cover acts of kindness or charity that go beyond strict obligation. In this document, beneficence is understood in a stronger sense, as an obligation. Two general rules have been formulated as complementary expressions of beneficent actions in this sense: (1) do not harm and (2) maximize possible benefits and minimize possible harms.

   The Hippocratic maxim "do no harm" has long been a fundamental principle of medical ethics. Claude Bernard extended it to the realm of research, saying that one should not injure one person regardless of the benefits that might come to others. However, even avoiding harm requires learning what is harmful; and, in the process of obtaining this information, persons may be exposed to risk of harm. Further, the Hippocratic Oath requires physicians to benefit their patients "according to their best judgment." Learning what will in fact benefit may require exposing persons to risk. The problem posed by these imperatives is to decide when it is justifiable to seek certain benefits despite the risks involved, and when the benefits should be foregone because of the risks.

   The obligations of beneficence affect both individual investigators and society at large, because they extend both to particular research projects and to the entire enterprise of research. In the case of particular projects, investigators and members of their institutions are obliged to give forethought to the maximization of benefits and the reduction of risk that might occur from the research investigation. In the case of scientific research in general, members of the larger society are obliged to recognize the longer term benefits and risks that may result from the improvement of knowledge and from the development of novel medical, psychotherapeutic, and social procedures.

The principle of beneficence often occupies a well-defined justifying role in many areas of research involving human subjects. An example is found in research involving children.

Effective ways of treating childhood diseases and fostering healthy development are benefits that serve to justify research involving children—even when individual research subjects are not direct beneficiaries. Research also makes it possible to avoid the harm that may result from the application of previously accepted routine practices that on closer investigation turn out to be dangerous. But the role of the principle of beneficence is not always so unambiguous. A difficult ethical problem remains, for example, about research that presents more than minimal risk without immediate prospect of direct benefit to the children involved. Some have argued that such research is inadmissible, while others have pointed out that this limit would rule out much research promising great benefit to children in the future. Here again, as with all hard cases, the different claims covered by the principle of beneficence may come into conflict and force difficult choices.

3. **Justice**—Who ought to receive the benefits of research and bear its burdens? This is a question of justice, in the sense of "fairness in distribution" or "what is deserved." An injustice occurs when some benefit to which a person is entitled is denied without good reason or when some burden is imposed unduly. Another way of conceiving the principle of justice is that equals ought to be treated equally. However, this statement requires explication. Who is equal and who is unequal? What considerations justify departure from equal distribution? Almost all commentators allow that distinctions based on experience, age, deprivation, competence, merit and position do sometimes constitute criteria justifying differential treatment for certain purposes. It is necessary, then, to explain in what respects people should be treated equally. There are several widely accepted formulations of just ways to distribute burdens and benefits. Each formulation mentions some relevant property on the basis of which burdens and benefits should be distributed. These formulations are (1) to each person an equal share, (2) to each person according to individual need, (3) to each person according to individual effort, (4) to each person according to societal contribution, and (5) to each person according to merit.

Questions of justice have long been associated with social practices such as punishment, taxation and political representation. Until recently these questions have not generally been associated with scientific research. However, they are foreshadowed even in the earliest reflections on the ethics of research involving human subjects. For example, during the 19th and early 20th centuries the burdens of serving as research subjects fell largely upon poor ward patients, while the benefits of improved medical care flowed primarily to private patients. Subsequently, the exploitation of unwilling prisoners as research subjects in Nazi concentration camps was condemned as a particularly flagrant injustice. In this country, in the 1940s, the Tuskegee syphilis study used disadvantaged, rural black men to study the untreated course of a disease that is by no means confined to that population. These subjects were deprived of demonstrably effective treatment in order not to interrupt the project, long after such treatment became generally available.

Against this historical background, it can be seen how conceptions of justice are relevant to research involving human subjects. For example, the selection of research subjects needs to be scrutinized in order to determine whether some classes (for example, welfare patients, particular racial and ethnic minorities, or persons confined to institutions) are being systematically selected simply because of their easy availability, their compromised position, or their manipulability, rather than for reasons directly related to the problem being studied. Finally, whenever research supported by public funds leads to the development of therapeutic devices and procedures, justice demands both that these not provide advantages only to those who can afford them and that such research should not unduly involve persons from groups unlikely to be among the beneficiaries of subsequent applications of the research.

PART C: APPLICATIONS

Applications of the general principles to the conduct of research leads to consideration of the following requirements: informed consent, risk/benefit assessment, and the selection of subjects of research.

1. **Informed Consent**—Respect for persons requires that subjects, to the degree that they are capable, be given the opportunity to choose what shall or shall not happen to them. This opportunity is provided when adequate standards for informed consent are satisfied.

   While the importance of informed consent is unquestioned, controversy prevails over the nature and possibility of an informed consent. Nonetheless, there is widespread agreement that the consent process can be analyzed as containing three elements: information, comprehension and voluntariness.

   Information. Most codes of research establish specific items for disclosure intended to assure that subjects are given sufficient information. These items generally include: the research procedure, their purposes, risks and anticipated benefits, alternative procedures (where therapy is involved), and a statement offering the subject the opportunity to ask questions and to withdraw at any time from the research. Additional items have been proposed, including how subjects are selected, the person responsible for the research, etc.

   However, a simple listing of items does not answer the question of what the standard should be for judging how much and what sort of information should be provided. One standard frequently invoked in medical practice, namely the information commonly provided by practitioners in the field or in the locale, is inadequate since research takes place precisely when a common understanding does not exist. Another standard, currently popular in malpractice law, requires the practitioner to reveal the information that reasonable persons would wish to know in order to make a decision regarding their care. This, too, seems insufficient since the research subject, being in essence a volunteer, may wish to know considerably more about risks gratuitously undertaken than do patients who deliver themselves into the hand of a clinician for needed care.

It may be that a standard of "the reasonable volunteer" should be proposed: the extent and nature of information should be such that persons, knowing that the procedure is neither necessary for their care nor perhaps fully understood, can decide whether they wish to participate in the furthering of knowledge. Even when some direct benefit to them is anticipated, the subjects should understand clearly the range of risk and the voluntary nature of participation.

A special problem of consent arises where informing subjects of some pertinent aspect of the research is likely to impair the validity of the research. In many cases, it is sufficient to indicate to subjects that they are being invited to participate in research of which some features will not be revealed until the research is concluded. In all cases of research involving incomplete disclosure, such research is justified only if it is clear that (1) incomplete disclosure is truly necessary to accomplish the goals of the research, (2) there are no undisclosed risks to subjects that are more than minimal, and (3) there is an adequate plan for debriefing subjects, when appropriate, and for dissemination of research results to them. Information about risks should never be withheld for the purpose of eliciting the cooperation of subjects, and truthful answers should always be given to direct questions about the research. Care should be taken to distinguish cases in which disclosure would destroy or invalidate the research from cases in which disclosure would simply inconvenience the investigator.

Comprehension. The manner and context in which information is conveyed is as important as the information itself. For example, presenting information in a disorganized and rapid fashion, allowing too little time for consideration or curtailing opportunities for questioning, all may adversely affect a subject's ability to make an informed choice.

Because the subject's ability to understand is a function of intelligence, rationality, maturity and language, it is necessary to adapt the presentation of the information to the subject's capacities. Investigators are responsible for ascertaining that the subject has comprehended the information. While there is always an obligation to ascertain that the information about risk to subjects is complete and adequately comprehended, when the risks are more serious, that obligation increases. On occasion, it may be suitable to give some oral or written tests of comprehension.

Special provision may need to be made when comprehension is severely limited—for example, by conditions of immaturity or mental disability. Each class of subjects that one might consider as incompetent (for example, infants and young children, mentally disable patients, the terminally ill and the comatose) should be considered on its own terms. Even for these persons, however, respect requires giving them the opportunity to choose to the extent they are able, whether or not to participate in research. The objections of these subjects to involvement should be honored, unless the research entails providing them a therapy unavailable elsewhere. Respect for persons also requires seeking the permission of other parties in order to protect the subjects from harm. Such persons are thus respected both by acknowledging their own wishes and by the use of third parties to protect them from harm.

The third parties chosen should be those who are most likely to understand the incompetent subject's situation and to act in that person's best interest. The person authorized to act on behalf of the subject should be given an opportunity to observe the research as it proceeds in order to be able to withdraw the subject from the research, if such action appears in the subject's best interest.

Voluntariness. An agreement to participate in research constitutes a valid consent only if voluntarily given. This element of informed consent requires conditions free of coercion and undue influence. Coercion occurs when an overt threat of harm is intentionally presented by one person to another in order to obtain compliance. Undue influence, by contrast, occurs through an offer of an excessive, unwarranted, inappropriate or improper reward or other overture in order to obtain compliance. Also, inducements that would ordinarily be acceptable may become undue influences if the subject is especially vulnerable.

Unjustifiable pressures usually occur when persons in positions of authority or commanding influence—especially where possible sanctions are involved—urge a course of action for a subject. A continuum of such influencing factors exists, however, and it is impossible to state precisely where justifiable persuasion ends and undue influence begins. But undue influence would include actions such as manipulating a person's choice through the controlling influence of a close relative and threatening to withdraw health services to which an individual would otherwise be entitled.

2. **Assessment of Risks and Benefits**—The assessment of risks and benefits requires a careful arrayal of relevant data, including, in some cases, alternative ways of obtaining the benefits sought in the research. Thus, the assessment presents both an opportunity and a responsibility to gather systematic and comprehensive information about proposed research. For the investigator, it is a means to examine whether the proposed research is properly designed. For a review committee, it is a method for determining whether the risks that will be presented to subjects are justified. For prospective subjects, the assessment will assist the determination whether or not to participate.

   The Nature and Scope of Risks and Benefits. The requirement that research be justified on the basis of a favorable risk/benefit assessment bears a close relation to the principle of beneficence, just as the moral requirement that informed consent be obtained is derived primarily from the principle of respect for persons. The term "risk" refers to a possibility that harm may occur.

However, when expressions such as "small risk" or "high risk" are used, they usually refer (often ambiguously) both to the chance (probability) of experiencing a harm and the severity (magnitude) of the envisioned harm.

The term "benefit" is used in the research context to refer to something of positive value related to health or welfare. Unlike, "risk," "benefit" is not a term that expresses probabilities. Risk is properly contrasted to probability of benefits, and benefits are properly contrasted with harms rather than risks of harm. Accordingly, so-called risk/benefit assessments are concerned with the probabilities and magnitudes of possible harm and anticipated benefits. Many kinds of possible harms and benefits need to be taken into account. There are, for example, risks of psychological harm, physical harm, legal harm, social harm and economic harm and the corresponding benefits. While the most likely types of harms to research subjects are those of psychological or physical pain or injury, other possible kinds should not be overlooked.

Risks and benefits of research may affect the individual subjects, the families of the individual subjects, and society at large (or special groups of subjects in society). Previous codes and Federal regulations have required that risks to subjects be outweighed by the sum of both the anticipated benefit to the subject, if any, and the anticipated benefit to society in the form of knowledge to be gained from the research. In balancing these different elements, the risks and benefits affecting the immediate research subject will normally carry special weight. On the other hand, interests other than those of the subject may on some occasions be sufficient by themselves to justify the risks involved in the research, so long as the subjects' rights have been protected. Beneficence thus requires that we protect against risk of harm to subjects and also that we be concerned about the loss of the substantial benefits that might be gained from research.

The Systematic Assessment of Risks and Benefits. It is commonly said that benefits and risks must be "balanced" and shown to be "in a favorable ratio." The metaphorical character of these terms draws attention to the difficulty of making precise judgments. Only on rare occasions will quantitative techniques be available for the scrutiny of research protocols. However, the idea of systematic, nonarbitrary analysis of risks and benefits should be emulated insofar as possible.

This ideal requires those making decisions about the justifiability of research to be thorough in the accumulation and assessment of information about all aspects of the research, and to consider alternatives systematically. This procedure renders the assessment of research more rigorous and precise, while making communication between review board members and investigators less subject to misinterpretation, misinformation and conflicting judgments. Thus, there should first be a determination of the validity of the presuppositions of the research; then the nature, probability and magnitude of risk should be distinguished with as much clarity as possible. The method of ascertaining risks should be explicit, especially where there is no alternative to the use of such vague categories as small or slight risk. It should also be determined whether an investigator's estimates of the probability of harm or benefits are reasonable, as judged by known facts or other available studies.

Finally, assessment of the justifiability of research should reflect at least the following considerations: (i) Brutal or inhumane treatment of human subjects is never morally justified. (ii) Risks should be reduced to those necessary to achieve the research objective. It should be determined whether it is in fact necessary to use human subjects at all. Risk can perhaps never be entirely eliminated, but it can often be reduced by careful attention to alternative procedures. (iii) When research involves significant risk of serious impairment, review committees should be extraordinarily insistent on the justification of the risk (looking usually to the likelihood of benefit to the subject—or, in some rare cases, to the manifest voluntariness of the participation). (iv) When vulnerable populations are involved in research, the appropriateness of involving them should itself be demonstrated. A number of variables go into such judgments, including the nature and degree of risk, the condition of the particular population involved, and the nature and level of the anticipated benefits. (v) Relevant risks and benefits must be thoroughly arrayed in documents and procedures used in the informed consent process.

3. **Selection of Subjects**—Just as the principle of respect for persons finds expression in the requirements for consent, and the principle of beneficence in risk/benefit assessment, the principle of justice gives rise to moral requirements that there be fair procedures and outcomes in the selection of research subjects.

   Justice is relevant to the selection of subjects of research at two levels: the social and the individual. Individual justice in the selection of subjects would require that researchers exhibit fairness: thus, they should not offer potentially beneficial research only to some patients who are in their favor or select only "undesirable" persons for risky research. Social justice requires that distinction be drawn between classes of subjects that ought, and ought not, to participate in any particular kind of research, based on the ability of members of that class to bear burdens and on the appropriateness of placing further burdens on already burdened persons. Thus, it can be considered a matter of social justice that there is an order of preference in the selection of classes of subjects (for example, adults before children) and that some classes of potential subjects (for example, the institutionalized mentally infirm or prisoners) may be involved as research subjects, if at all, only on certain conditions.

   Injustice may appear in the selection of subjects, even if individual subjects are selected fairly by investigators and treated fairly in the course of research. Thus injustice arises from social, racial, sexual and cultural biases institutionalized in society. Thus, even if individual researchers are treating their research subjects fairly, and even if IRBs are taking care to assure that subjects are selected fairly within a particular institution, unjust social patterns may nevertheless appear in the overall distribution of the burdens and benefits of research. Although individual institutions or investigators may not be able to resolve a problem that is pervasive in their social setting, they can consider distributive justice in selecting research subjects.

Some populations, especially institutionalized ones, are already burdened in many ways by their infirmities and environments. When research is proposed that involves risks and does not include a therapeutic component, other less burdened classes of persons should be called upon first to accept these risks of research, except where the research is directly related to the specific conditions of the class involved. Also, even though public funds for research may often flow in the same directions as public funds for health care, it seems unfair that populations dependent on public health care constitute a pool of preferred research subjects if more advantaged populations are likely to be the recipients of the benefits.

One special instance of injustice results from the involvement of vulnerable subjects. Certain groups, such as racial minorities, the economically disadvantaged, the very sick, and the institutionalized may continually be sought as research subjects, owing to their ready availability in settings where research is conducted. Given their dependent status and their frequently compromised capacity for free consent, they should be protected against the danger of being involved in research solely for administrative convenience, or because they are easy to manipulate as a result of their illness or socioeconomic condition.

[1] Since 1945, various codes for the proper and responsible conduct of human experimentation in medical research have been adopted by different organizations. The best known of these codes are the Nuremberg Code of 1947, the Helsinki Declaration of 1964 (revised in 1975), and the 1971 Guidelines (codified into Federal Regulations in 1974) issued by the U.S. Department of Health, Education, and Welfare Codes for the conduct of social and behavioral research have also been adopted, the best known being that of the American Psychological Association, published in 1973.

[2] Although practice usually involves interventions designed solely to enhance the well-being of a particular individual, interventions are sometimes applied to one individual for the enhancement of the well-being of another (for example, blood donation, skin grafts, organ transplants) or an intervention may have the dual purpose of enhancing the well-being of a particular individual, and, at the same time, providing some benefit to others (for example, vaccination, which protects both the person who is vaccinated and society generally). The fact that some forms of practice have elements other than immediate benefit to the individual receiving an intervention, however, should not confuse the general distinction between research and practice. Even when a procedure applied in practice may benefit some other person, it remains an intervention designed to enhance the well-being of a particular individual or groups of individuals; thus, it is practice and need not be reviewed as research.

[3] Because the problems related to social experimentation may differ substantially from those of biomedical and behavioral research, the Commission specifically declines to make any policy determination regarding such research at this time. Rather, the Commission believes that the problem ought to be addressed by one of its successor bodies.

Appendix B: Templates and Tools

SOP TITLE: PROTOCOL FEASIBILITY, REVIEW
SOP #: 202
Approval Date: 09 September 2016
Site Implementation Date: 02 October 2016
Prepared by/Approved by: C. Wells/J. Stevens
Last Revised: 09 September 2016
Effective Date: 02 October 2016

OBJECTIVE:

The objective of this Standard Operating Procedure (SOP) is to describe the procedures for assessing study feasibility for protocols to be conducted at Wells Clinical Research Site.

RESPONSIBILITIES:

- This SOP applies to the protocol review committee, principal investigator (PI), and other ad hoc research personnel who are responsible for making decisions about participation in clinical research at Wells Clinical Research Site.

PROCEDURES:

Scientific/Ethical Feasibility

- The PI and the protocol review committee review the protocol for:
    - Clinical importance to Wells Clinical Research Site patients
    - Scientific merit
    - Benefits and risks associated with the protocol
    - Vulnerable populations
    - Consistency with organizational mission

Regulatory Feasibility

- The protocol review committee reviews the protocol to determine whether anything that is required in the protocol may be problematic. As a second step, the research compliance officer reviews the protocol to determine any risks.

Operational Feasibility

- The PI and the protocol review committee review the protocol for:
  - Availability of research and specialty personnel to conduct the study procedures
  - Availability of appropriate recruitment population
  - Operational complexity of the protocol
  - Conflicting studies

Financial and Contract Feasibility

- The research manager and research accountant review the study to determine costs, staff time, etc.
- The research accountant compares costs vs. sponsor budget to determine if sponsor budget is reasonable and covers all costs.

REFERENCES:

- 21 CFR 56.109 IRB review of research
- 21 CFR 56.111 Criteria for IRB approval of research
- 21 CFR 312.21 Phases of an investigation
- 21 CFR 312.23 IND content and format
- 21 CFR 312.60 General responsibilities of investigators

ASSOCIATED SOPS:

- SOP 301 Industry-sponsored protocol budget review
- SOP 302 Negotiation of industry-sponsored protocols
- SOP 303 Industry-sponsored protocol contract review
- SOP 110 Protocol review committee roles and responsibilities
- SOP 111 Scientific review of industry-sponsored research protocols
- SOP 112 Ethical review of industry-sponsored research protocols
- SOP 113 Regulatory review of industry-sponsored research protocols

Adverse Event Form

STUDY NAME

Site Number: ____________

Pt_ID: ____________

Has the participant had any Adverse Events during this study? ☐ Yes ☐ No (*If yes, please list all Adverse Events below*)

Severity	Study Intervention Relationship	Action Taken Regarding Study Intervention	Outcome of AE	Expected	Serious
1 = Mild 2 = Moderate 3 = Severe	1 = Definitely related 2 = Possibly related 3 = Not related	1 = None 2 = Discontinued permanently 3 = Discontinued temporarily 4 = Reduced dose 5 = Increased dose 6 = Delayed dose	1 = Resolved, no sequel 2 = AE still present - no treatment 3 = AE still present - being treated 4 = Residual effects present - not treated 5 = Residual effects present - treated 6 = Death 7 = Unknown	1 = Yes 2 = No	1 = Yes 2 = No (If yes, complete SAE form)

Adverse Event	Start Date	Stop Date	Severity	Relationship to Study Treatment	Action Taken	Outcome of AE	Expected?	Serious Adverse Event?	Initials
1.									
2.									
3.									

Adverse Event

Form Version 1.0

Concomitant Medications Form

STUDY NAME	
Site Name:______________________ Pt_ID:______________________	This form is cumulative and may be used to capture concomitant medications of a single participant throughout the study.

At end of study only: Check this box if participant took no concomitant medication ☐ None

Medication	Indication	Dosage	Start Date	Stop Date	Ongoing?
					☐
					☐
					☐
					☐
					☐
					☐
					☐
					☐
					☐
					☐
					☐
					☐
					☐
					☐
					☐
					☐
					☐
					☐
					☐
					☐
					☐
					☐
					☐

Concomitant Medication Form Page __ of __ Version 2.0

MASTER INVESTIGATIONAL PRODUCTACCOUNTABILITY LOG

PROTOCOL TITLE:		
PRINCIPAL INVESTIGATOR:	SITE:	INITIATION DATE:
TYPE OF INVESTIGATIONAL PRODUCT:		

Number of IP received	Date Received	Batch Number	Expiry Date	Subject Number	Number of IP dispensed	Balance Number of IP	Signature	Did subject return any IP? (Yes or No) Comment (*if any*)

**Please use a new log once you have completed entry in all rows in this log. PI to verify and acknowledge all entry in this IP accountability log.*

Signature:
Date:

Version Date X.X. Date: DD/MMM/YYYY

Page ___ of ____

Sample Checklist for Clinical Trial Feasibility at Research Site

Scientific Review	
	Is the study within the scope of practice of the investigator?
	Does the study have scientific merit?
	Do the objectives and scientific methodology of the protocol support the purpose of the study?
	Are the inclusion/exclusion criteria appropriate for the study?
	Do the preclinical studies support the purpose and conduct of the clinical study?
	Is the information related to adverse event reporting aligned with the conduct of the study?
	Is the method of administration for the investigational product appropriate?
	Are the references and the background for the protocol supported scientifically?
Regulatory Review	
	Does the protocol meet institutional/local/state requirements/regulations?
	Does the protocol include all the ingredients required in 21 C.F.R. Part 312.23 (a)(6)(iii)?
Ethical Review	
	Is the protocol ethical? (How will the IRB view the study?)
	Are there concerns in regard to subject autonomy–is the informed consent process appropriate?
	Is the risk/benefit ratio appropriate for the protocol?
	Is the study population target appropriate?
Resource Review	
	Population
	Is there access to target subject population?
	Is enrollment goal and timeline realistic?
	Will enrollment compete with other studies seeking same subject population?
	Are the inclusion/exclusion criteria too restrictive? (Consider number of screen failures)
	Are vulnerable populations involved in the study? (Consider special consent issues)
	Staff
	Is the staff qualified?
	Is training available?
	Does the PI have adequate time and training to devote to the protocol?
	Is it anticipated that there will be numerous queries?
	What is the monitoring plan–will it require additional coordinator time?
	Facilities
	Is adequate clinic and office space available?
	Is any special equipment required?
	Is access to emergency rescue equipment necessary?
	Supplies
	What will the sponsor supply (CRFs, source documents, electronic consent template, packaged lab kits, pre-paid shipping, etc.)?
	Will electronic or remote data capture be used? If so, will sponsor provide hardware and training?
	Budget
	Is preliminary budget adequate?
	Will sponsor pay for items such as protocol amendments, reconsent of subjects, unanticipated monitoring, etc.?
	Will sponsor pay for an adequate number of screen failures?

Note: This is a sample checklist and is not fully comprehensive of all elements that should be considered in determining if a study is feasible for a site.

Research Site Initiation Checklist

- ✓ Complete and submit the FDA Form 1572 to the Sponsor.
- ✓ Submit current copies of the investigator(s) CVs, medical license, and proof of insurance to the Sponsor.
- ✓ Submit requested documentation of staff training and qualifications to the Sponsor.
- ✓ Install the electronic data capture system and train, conduct and document training of staff.
- ✓ Investigator must complete and submit the Delegation of Authority form to the sponsor.
- ✓ Obtain IRB approval of the site-specific informed consent.
- ✓ Obtain completed and signed financial disclosures forms from applicable staff and submit to the Sponsor.
- ✓ Obtain completed and signed confidentiality agreements from applicable staff and submit to the Sponsor.
- ✓ Make sure study staff has completed any required research training and study-specific training, that is has been documented and submitted to the Sponsor.
- ✓ The investigator must sign a document that he/she has reviewed and understands the protocol and the investigator brochure.
- ✓ The study budget must be approved and signed by appropriate parties.
- ✓ The clinical trial agreement must be approved and signed by appropriate parties.
- ✓ Site has received all study supplies promised by the sponsor.
- ✓ Site has received, inventoried and stored investigational product.
- ✓ Site has designed a study recruitment plan and that meets the recruitment terms outlined in the clinical trial agreement.

Appendix C: Clinical Trial Resources

This appendix provides readers with some of the organizations and businesses that provide tools, resources, and training for clinical trial operations. The inclusion of a resource in this appendix is not an endorsement of the resource by the authors. This is not an all-inclusive list of the resources available.

BUDGETS AND CONTRACTS

Pfeiffer, J.P., & Windschiegl, M. (2016). *Managing clinical trial budgets and contracts*. Georgia: LAD Customer Publisher.

Model Agreements and Guidelines International (MAGI) (www.magiworld.org/)

Mission

Model Agreements & Guidelines International (MAGI) is streamlining clinical research by standardizing best practices for clinical operations, business and regulatory compliance. MAGI has the following contract templates available:

- Clinical Trial Agreement Template
- Clinical Trial Handbook
- Budget Template

PFS Clinical (http://pfsclinical.com)

Over several years of providing services to research institutions, we've come to recognize that no two institutions are the same. Accordingly, our solutions are designed to meet your unique needs. Upon partnering with PFS Clinical, the relationship begins with a discovery process during which we thoroughly review your clinical research portfolio to determine how we can best help you.

Study-specific services

- Coverage Analysis
- Initial Regulatory Submission
- Internal Budget Builds
- Contract Redlines
- Contract/Budget Negotiations
- Financial Insight & Management

CLINICAL TRIAL MANAGEMENT SYSTEMS

Alpha (http://alphaclinicalsystems.com)

Alpha Clinical System's, Clinical Trials Management System (ez-CTMS) is a modular, interoperable and standards-based software application designed to meet needs of a single-site or multi-site global clinical trials of CRO's or Pharmaceutical clients.

The flexible framework and exceptional breadth of functionality of ez-CTMS benefits the user in creating, planning, managing and monitoring clinical studies along with the budget and finance management.

Bioclinica (http://www.bioclinica.com/)

eSource® Express equips study teams with easy-to-use tools for quick and efficient data collection right at the source (yes, this is eSource!) on any device or desktop. A built-in ePRO tool makes it convenient for subjects to enter required data between visits. Express provides all of the essential tools—each tailored to the individual user. Express makes all aspects of the clinical trial process easier, faster, and more efficient—with clean data right at the point of entry.

OnPoint CTMS is a powerful end-to-end clinical trials management solution that brings control, efficiency, and quality data to every study. This web-based CTMS works on the go, letting you view and manage real-time operational performance—wherever and whenever.

Covance CTMS (http://www.covance.com)

Xcellerate® Trial Management, our advanced Clinical Trial Operating Platform includes our Electronic Trial Master File (eTMF) and Clinical Trial Management System (CTMS) Solutions.

Clinical trials often involve multiple trial sites, team members across different time zones and multiple complex documents. As a result, managing the authoring, maintenance and access of all study documentation presents a significant challenge.

To advance your clinical trial quality and enhance your ability to obtain high-quality, audit-ready data and documentation, you need a partner who can provide the tools and technology to streamline data integration, efficiently manage your study and accelerate your results.

The Covance Advanced Clinical Trial Operating Platform—which includes our Electronic Trial Master File (eTMF) and Clinical Trial Management System (CTMS) solutions—will empower you and your team to monitor the progress of your studies continuously and identify potential problems early—allowing the right people to make key decisions at the right time.

ClickCTMs (http://www.huronconsultinggroup.com)

Huron's Click® CTMS is a flexible, comprehensive clinical trials management system that helps you run your research enterprise like the business it is. Budget, code, and bill clinical trials correctly, consistently. Improve cash flow through timely sponsor billing and collections. Improve contract decision making with margin visibility. Reduce compliance risk through IRB and COI integration.

Configurable to match your forms and workflow processes, the Click software integrates seamlessly with existing HR, financial, and EMR systems.

Trial Interactive—eTMF (http://www.trialinteractive.com)

As life sciences companies move to semi-virtual environments keeping only the most vital functions in house, more and more Trial Master Files are being converted to an eTMF format.

Trial Interactive provides companies with the ability to set up eTMFs in a matter of hours and can even be organized based on the file structure which is currently in place, allowing access to internal users as well as any partners with high level security access. This significantly reduces the costs of corporate IT support, as well as the challenges of tracking this information with hard copies. Trial Interactive enables you to host your eTMFs online with drag and drop functionality, allowing regulatory administrators to rapidly reorganize content. This accelerates regulatory processes, while providing documentation access to key personnel.

ClinPlus (http://www.clinplus.com)

The ClinPlus® eClinical Platform combines our electronic data capture and clinical trial management system software into one, easy-to-use platform that maximizes productivity while minimizing risk. Through superior data and trial management, this innovative platform keeps clinical trials running efficiently.

eResearch (http://velos.com)

Velos eResearch is the most comprehensive and adaptable clinical research management suite (CRMS) available for automating all administrative, financial, and research activities.

Designed to promote productivity and efficiency, Velos eResearch simplifies the management of the entire clinical research portfolio by linking study status, patient enrollment, calendars, budgets, electronic data capture and more.

Florence Healthcare (http://florencehc.com/)

Florence Healthcare, a startup team led by Emory, Microsoft, and Airwatch execs, invented new a way to gather records automatically from trial sites. This gives sponsors access to source documents for setup, remote monitoring and adverse events that are often behind locked EHRs or on paper—allowing vastly improved site compliance and faster closeout.

Fortress (https://www.fortressmedical.com)

As a comprehensive clinical trial management system, Clindex drives efficiencies and results across the entire clinical trial process, from site selection through study close-out.

CTMS software features include IRB/EC, regulatory, and protocol tracking; visit scheduling and compliance; enrollment tracking; product management; comprehensive e-monitoring; site payments; document management system; plus the ability to add features unique to your needs.

Unlike some Clinical Trial Management Systems that are created by connecting existing CTMS and EDC systems, Clindex is a single system, offering you all the functionality you need in a single eClinical solution.

Forte (http://forteresearch.com)

Oncore® Enterprise—Developed through collaboration with leading research organizations, the OnCore® Enterprise Research system provides proven functionality for supporting efficient processes at academic medical centers, cancer centers, and health care systems.

Allegro—cloud-based clinical trial management system that manages operational data for small to mid-size sites.

Medidata Clinical Trial System (https://www.mdsol.com)

It's configurable. Cloud technology allows Medidata CTMS to offer the same power to every size organization—from top 10 pharmas to startups and CROs—with no IT expertise required. Simple configuration tailors CTMS to the unique requirements of your study.

It's agile. Agile development methodology powers regular functional enhancements to CTMS and the cloud puts those enhancements quickly into your hands. So CTMS always keeps pace with your clinical research processes.

It's open. Don't battle costly integrations or fragmented data. Medidata CTMS's open architecture seamlessly exchanges information with other systems—pre-populating data, triggering events and eliminating duplicate work. There's even a configurable connection to Medidata Rave right in the Medidata Clinical Cloud.

DOCUMENT TRANSLATION SERVICES

GTS (https://www.gts-translation.com/)

GTS is a translation company that has been providing high quality, professional translation services for over 15 years. At GTS, the quality of our translation service is our number one concern! Our quality system is compliant with two recognized quality standards and is audited annually by an international certification body. All of our translations are reviewed before delivery to ensure that you are getting a good translation.

We use an entirely human translation process, supported by a stringent quality policy. Using the latest in translation software technology, our workflows are streamlined to translate your texts as quickly and as accurately as possible. Our quality system is certified as compliant with both the EN 15038 and ISO 9001:2008 quality standard. If you require official certified translations of documents, we can provide them for you. Many of our translators are certified by accredited translation organizations. We can also provide court-approved sworn translations for many countries.

Language Scientific (www.languagescientific.com)

Language Scientific is a US-based global translation company specializing in Clinical and Scientific translation services for the Life Sciences industry. Our focus is on helping companies involved with clinical trials, such as CROs and Pharmaceutical companies, make their global operations run seamlessly. We deliver fast, high quality translations by customizing our services, processes and software platforms to fit the exact needs of our clients.

The company translates over 40 million words a year for Clinical Trials. Clinical Translation services makes up our largest business segment, accounting for 83% of our revenue. The largest portion of our Clinical business is in support of global clinical trials, where we translate study start up documents such as protocols, investigator brochures, and informed consent forms, conduct linguistic validation and cognitive debriefing on PRO instruments and translate adverse event reports and case source documents for endpoint adjudication and clinical safety initiatives.

Language Scientific translates and manages a diverse set of clinical documents including the following:

- Adverse Event Source Documents
- Case Report Forms (CRFs)
- Clinical Trial Agreements (CTAs)
- Data Sheets
- Drug Registration Documentation
- Endpoint Adjudication Documents
- Informed Consent Forms
- Insert Leaflets
- Investigator Brochures
- Marketing Authorization Applications (MAAs)
- New Drug Applications (NDAs)
- Package Inserts and Labels
- Patient Diaries
- Patient Recruitment Materials
- Patient Reported Outcome Measures (PROMs)
- Patient Source Documents, Admission and Discharge, Labs
- Pharmacological Studies
- QoL Scales
- Rater Training
- Regulatory Documents

- Site Operations Manuals
- Study Protocols
- Toxicology Reports
- Trial Master File (TMF)

Novalins (http://www.novalins.com/regulatory-affairs-translation/)

Novalins is a clinical research translation specialist dedicated to deliver the highest standards in linguistic products to clients worldwide, while always being on-time and within budget.

Our years of experience in the clinical trials field enables us to provide industry-standard expertise and to adapt our processes according to the specific needs of each document at every stage of the product development and registration process:

- Preclinical studies;
- Phases 0 to IV;
- Surveys & Drug testing;
- Dossiers for regulatory approval.

As clinical trials are increasingly conducted across national and language borders, there is a heightened need for accurate and efficient translation of the documents and communication used in clinical trials. Getting it right the first time is fundamental because so much invested time, money and reputation is at stake.

Moravia (http://www.moravia.com/en/services/life-sciences/)

Moravia is a leading provider of translation, localization, and testing services. Our globalization solutions enable companies to enter global markets with high quality localized products and services that meet the language and functionality requirements of customers in any locale.

Helping medical device manufacturers, pharmaceuticals manufacturers and Contract Research Organizations (CROs) conducting clinical trials by translating associated materials such as Informed Consent Forms, study protocols, questionnaires and instructions, patient diaries and more.

GOVERNMENT ORGANIZATIONS

Centers for Disease Control and Prevention (CDC) (http://www.cdc.gov/)

The CDC is one of the major operating components of the Department of Health and Human Services.

CDC works 24/7 to protect Americans from health, safety and security threats, both foreign and in the U.S. Whether diseases start at home or abroad, are chronic or acute, curable or preventable, human error or deliberate attack, CDC fights disease and supports communities and citizens to do the same.

CDC increases the health security of our nation. As the nation's health protection agency, CDC saves lives and protects people from health threats. To accomplish our mission, CDC conducts critical science and provides health information that protects our nation against expensive and dangerous health threats, and responds when these arise.

CDC's Role

- Detecting and responding to new and emerging health threats
- Tackling the biggest health problems causing death and disability for Americans
- Putting science and advanced technology into action to prevent disease
- Promoting healthy and safe behaviors, communities and environment
- Developing leaders and training the public health workforce, including disease detectives
- Taking the health pulse of our nation

HTTPS://CLINICALTRIALS.GOV/ IS A SERVICE OF THE U.S. NATIONAL INSTITUTES OF HEALTH

ClinicalTrials.gov is a registry and results database of publicly and privately supported clinical studies of human participants conducted around the world.

ClinicalTrials.gov is a Web-based resource that provides patients, their family members, health care professionals, researchers, and the public with easy access to information on publicly and privately supported clinical studies on a wide range of diseases and conditions. The Web site is maintained by the National Library of Medicine (NLM) at the National Institutes of Health (NIH). Information on ClinicalTrials.gov is provided and updated by the sponsor or principal investigator of the clinical study. Studies are generally submitted to the Web site (that is, registered) when they begin, and the information on the site is updated throughout the study. In some cases, results of the study are submitted after the study ends. This Web site and database of clinical studies is commonly referred to as a "registry and results database."

ClinicalTrials.gov contains information about medical studies in human volunteers. Most of the records on ClinicalTrials.gov describe clinical trials (also called interventional studies). A clinical trial is a research study in which human volunteers are assigned to interventions (for example, a medical product, behavior, or procedure) based on a protocol (or plan) and are then evaluated for effects on biomedical or health outcomes. ClinicalTrials.gov also contains records describing and programs providing access to investigational drugs outside of clinical trials (expanded access). Studies listed in the database are conducted in all 50 States and in 192 countries.

US DEPARTMENT OF HEALTH AND HUMAN SERVICES (HHS) (HTTP://WWW.HHS.GOV/ABOUT/INDEX.HTML)

It is the mission of the U.S. Department of Health & Human Services (HHS) to enhance and protect the health and well-being of all Americans. We fulfill that mission by providing for effective health and human services and fostering advances in medicine, public health, and social services.

HHS has 11 operating divisions, including eight agencies in the U.S. Public Health Service and three human services agencies. These divisions administer a wide variety of health and human services and conduct life-saving research for the nation, protecting and serving all Americans.

The Office of the Secretary (OS), HHS's chief policy officer and general manager, administers and oversees the organization, its programs, and its activities. The Deputy Secretary and a number of Assistant Secretaries and Offices support OS.

US Food and Drug Administration (FDA) (http://www.fda.gov/)

FDA is responsible for protecting the public health by assuring the safety, efficacy and security of human and veterinary drugs, biological products, medical devices, our nation's food supply, cosmetics, and products that emit radiation.

FDA is also responsible for advancing the public health by helping to speed innovations that make medicines more effective, safer, and more affordable and by helping the public get the accurate, science-based information they need to use medicines and foods to maintain and improve their health. FDA also has responsibility for regulating the manufacturing, marketing and distribution of tobacco products to protect the public health and to reduce tobacco use by minors.

Finally, FDA plays a significant role in the Nation's counterterrorism capability. FDA fulfills this responsibility by ensuring the security of the food supply and by fostering development of medical products to respond to deliberate and naturally emerging public health threats.

National Institutes of Health (NIH) (https://www.nih.gov/about-nih)

The National Institutes of Health (NIH), a part of the U.S. Department of Health and Human Services, is the nation's medical research agency—making important discoveries that improve health and save lives.

A part of the U.S. Department of Health and Human Services, NIH is the largest biomedical research agency in the world.

NIH's mission is to seek fundamental knowledge about the nature and behavior of living systems and the application of that knowledge to enhance health, lengthen life, and reduce illness and disability.

The goals of the agency are:

- to foster fundamental creative discoveries, innovative research strategies, and their applications as a basis for ultimately protecting and improving health;
- to develop, maintain, and renew scientific human and physical resources that will ensure the Nation's capability to prevent disease;

- to expand the knowledge base in medical and associated sciences in order to enhance the Nation's economic well-being and ensure a continued high return on the public investment in research; and
- to exemplify and promote the highest level of scientific integrity, public accountability, and social responsibility in the conduct of science.

In realizing these goals, the NIH provides leadership and direction to programs designed to improve the health of the Nation by conducting and supporting research:

- in the causes, diagnosis, prevention, and cure of human diseases;
- in the processes of human growth and development;
- in the biological effects of environmental contaminants;
- in the understanding of mental, addictive and physical disorders; and
- in directing programs for the collection, dissemination, and exchange of information in medicine and health, including the development and support of medical libraries and the training of medical librarians and other health information specialists.

Office for Civil Rights (OCR) (http://www.hhs.gov/ocr/about-us/index.html)

Through the federal civil rights laws and Health Insurance Portability and Accountability Act (HIPAA) Privacy Rule, OCR protects your fundamental nondiscrimination and health information privacy rights by:

- Teaching health and social service workers about civil rights, health information privacy, and patient safety confidentiality laws
- Educating communities about civil rights and health information privacy rights
- Investigating civil rights, health information privacy, and patient safety confidentiality complaints to identify discrimination or violation of the law and take action to correct problems.

Office of Human Research Protections (OHRP) (http://www.hhs.gov/ohrp/)

The Office for Human Research Protections (OHRP) provides leadership in the protection of the rights, welfare, and wellbeing of human subjects involved in research conducted or supported by the U.S. Department of Health and Human Services (HHS). OHRP is part of the Office of the Assistant Secretary for Health in the Office of the Secretary of HHS.

OHRP provides clarification and guidance, develops educational programs and materials, maintains regulatory oversight, and provides advice on ethical and regulatory issues in biomedical and behavioral research. OHRP also supports the

Secretary's Advisory Committee on Human Research Protections (SACHRP), which advises the HHS Secretary on issues related to protecting human subjects in research.

INSTITUTIONAL REVIEW BOARDS

Chesapeake IRB (http://www.chesapeakeirb.com/about-us/)

Vision

To foster contributions to improve health and wellbeing throughout the world.

Mission

Partnering with sponsors, CROs, institutions and researchers to promote the highest level of human subject protection through scientifically and ethically sound research.

Continuously evolving our processes to exceed customer expectations while complying with regulations, legal, and ethical requirements.

Advocating, with peers, regulators and the public in the interest of industry standards to assure human research protections.

Our Core Values

Our core values are an extension of our commitment to address all matters related to the protection of human research subjects:

- Breakthrough Innovation
- Problem Solving
- Responsiveness
- Careful Listening

Chesapeake IRB brings together true knowledge leaders in the field of research and depth of expertise in IRB services for study initiation. Not merely providers of ethical review services, Chesapeake IRB provides a solid foundation on which stakeholders can confidently rely to solve problems and address all matters related to the protection of human research subjects.

Through careful listening and responsiveness, Chesapeake IRB facilitates the conduct of human research to the highest ethical standards and quality. We are experts at making strategic use of effective technology, and continually advance the cutting edge of techniques, processes, and perspectives in support of effective and efficient conduct of clinical research.

Integra Review IRB (http://www.integreview.com/)

IntegReview IRB is committed to protecting the rights and welfare of human subjects participating in research by:

- Striving to go above and beyond federal regulations
- Ensuring research is ethical and safe
- Providing education for all parties involved in research

IntegReview IRB is committed to maintaining superior customer service without compromising ethical values by:

- Providing rapid, quality service
- Assisting clients with ethical compliance
- Demonstrating respect
- Offering flexibility

Core Values

Ethical = Strive to do the right thing.
Be honest and accountable for actions.

Flexible = Value and accept change and growth.
Empower employees to submit new ideas and process improvements. Honor specific client requests.

Quality = Do it right the first time.
Provide training to ensure compliance with policies and procedures.

Team Work = Functioning together to achieve a common goal.
Establish effective communications between staff members.

Environment of Respect = Show respect for and trust in others.
Promote recognition of accomplishments through rewards. Provide "open-door" policy to encourage honest communication.

New England IRB (NEIRB) (http://neirb.com/)

New England IRB is an independent, central institutional review board for sponsors, CROs and individual researchers across North America. Our priority is to ensure the safety of human subjects in clinical trials, and we are committed to an ethical and thorough review process.

Our Boards are comprised of highly qualified members with significant experience and knowledge in the ethical, scientific and legal aspects of clinical trials.

New England IRB reviews:

- All phases of FDA regulated clinical trials:
  - Phase I–IV
  - Phase I/IIa hybrid studies
  - Patient reported outcomes studies
- Drug, device and biologic studies (including vaccine studies)
- Registry studies
- Peri-approval and post-approval studies
- Socio-behavioral and educational studies
- Requests for Exemption

Founded in 1988, New England IRB was one of the first central IRBs established to meet the ethical review needs of the clinical trials industry. For studies ranging from one site to several thousand, New England IRB is focused on the protection of human subjects, responsiveness and service.

We offer responsive review timelines while maintaining mechanisms to ensure a thorough, duly diligent, high quality review.

Our secure web portal and electronic systems are helpful tools that create efficiencies in submissions and notifications.

New England IRB has been audited by the United States Food and Drug Administration, and found to be in compliance with regulations. We are qualified to serve as the IRB for federally funded research.

Pearl IRB (http://www.pearlirb.com/)

Pearl IRB is an independent institutional review board. Our team is comprised of experts in the fields of medical practice, science, ethics and clinical research. We will ensure that a quality, timely review of proposed studies is conducted in the best interest of the patient, sponsor and research institution.

At Pearl IRB, we deliver superior central IRB review services that effectively balance the needs of human subjects, sponsors, and sites. Our team is comprised of experts in the fields of medical practice, science, ethics and clinical research. Our vision is to improve the clinical research process thereby delivering new therapeutics and diagnostics to patients sooner. We also offer consulting services to assist in the planning and execution of clinical research.

Quorum Review IRB (http://www.quorumreview.com/)

Quorum Review IRB assures the ethical integrity of clinical research. We help our partners advance the frontiers of medical innovation with greater confidence and control while sustaining the rights, dignity and safety of participants.

Like you, we're fully invested in supporting clinical trials that realize life-saving advancements. We are responsible for the efficiency and integrity of our ethical oversight of a trial. From early preparations through initial review, from site submissions to safety reporting, we have engineered Quorum's systems and processes for easy use. And we remain agile, ready to respond to your study's specific needs.

As one of the largest privately held IRBs, we have the resources and experience to help you navigate the ethical, legal and regulatory risks associated with complex, far reaching clinical studies.

Schulman IRB (http://www.sairb.com/)

Founded in 1983, Schulman IRB is the leading independent institutional review board dedicated to safeguarding the rights and welfare of clinical research participants.

We present the research communities of the U.S., Puerto Rico and Canada with a unique value proposition: Thorough, timely ethical reviews, conducted by a deeply experienced Board, backed by highly responsive customer service and proprietary, Part 11 compliant technology tools. By combining these factors, we both improve human subject protection and address the operational needs of those we work with.

In addition to reviews of all phases of research across all therapeutic areas, we offer contract research organizations, investigators and institutions consulting services (through a joint venture, Provision Research Compliance Services).

Fully accredited by the Association for the Accreditation of Human Research Protection Programs (AAHRPP), Schulman has an unparalleled clean audit history with the Food and Drug Administration (FDA).

SOLUTIONS IRB (HTTP://WWW.SOLUTIONSIRB.COM/)

Solutions IRB is a private Institutional Review Board (IRB). Our mission is to support researchers who need or wish to obtain IRB approval. We have a team of experienced reviewers ready to support researchers with IRB services including the submission of the IRB application, protecting all research participants (including vulnerable populations), and monitoring approved protocols. Our team provides quality and timely IRB review for quantitative, qualitative, action research and participatory action research studies. Solutions IRB branches are located in Arkansas, Arizona, and Minnesota.

Solutions IRB services include:

- Free no obligation IRB pre-review process to streamline the official review and identify the level of review required (non-human subjects, exempt, expedited or full)
- 24 hour turnaround on complete submissions of exempt and expedited reviews
- 48–72 hour turnaround on complete submissions of more than minimal risk studies
- Multiple IRB Board meetings each week
- Multi-center or single site reviews
- Online submission
- Continuing Review reminders
- Education and training services

Types of Reviews:

- FDA Regulated Studies
- OHRP Regulated Studies
- Device Studies
- Biologic Studies
- Observational/Registry Studies
- Social-Behavioral Studies
- Educational Studies

STERLING IRB (HTTP://WWW.STERLINGIRB.COM/)

Sterling IRB offers a comprehensive suite of services designed to help you work smarter, faster and more efficiently.

IRB Services

- Central and local IRB services
- Comprehensive review services in U.S. and Canada
- Experienced and diverse Board
- Daily Board Meetings
- Pre-review services for protocols
- Informed Consent development
- Certified translation services
- Review of research involving all vulnerable populations, including prisoners, cognitively impaired persons, pregnant women, and children

Client Services

- Dedicated CIP Certified Account Manager for study duration
- Study start-up teleconferences
- Timely turnaround
  - 2 business day turnaround time (Expedited Review) on average from date of submission
  - 5 business day turnaround time (Full Board Review) on average from date of submission
- Written notification within 24 hours of IRB Meeting
- Site review within 24 hours
- Site start-up status reports according to your requirements
- Continuing review reminder reports and status reports according to your requirements

Experience

- Phase I–IV studies in all therapeutic areas
- Single and Multi-Site studies
- Federally-funded studies
- Late phase studies—Post Market, Observational, Risk Evaluation and Mitigation Strategies ("REMS"), Registries
- Medical Device studies
- Biologic studies
- Nutraceutical studies
- Social-Behavioral studies
- Minimal Risk/Non-interventional studies
- Exemptions
- Non-Human Subject Research and Ethical Evaluations
- Knowledge, Attitudes, and Beliefs ("KABs")

Western IRB (WIRB) (https://www.wirb.com)

For more than 40 years, WIRB has been at the forefront of protecting the rights and welfare of human subjects. We provide in-depth regulatory expertise to support your development of research protocols and documentation.

We offer customized training and consulting about human protections, addressing research design through implementation. And, we deliver unparalleled review quality to ensure your research withstands scrutiny around the world.

Imagine one point of contact for all your needs. IRB review, biosafety, translation, international protocols—no matter how complicated your study, WIRB has the capacity and expertise to assist you. We are also the only independent IRB to offer additional safety services.

Our job is to simplify complexity. Responsive account representatives are ready to answer questions and help you successfully submit and track your protocol. Use our web portal to get step-by-step instructions, estimate turnaround times, track progress, and more.

We are there for subjects as well, with a live operator available any time of day and a WIRB staff physician always on call for inquiries about significant safety concerns or emergencies.

IRB MANAGEMENT SOFTWARE

iMedRIS (https://imedris.com/Modules/IRB-Software)

From creating proposals to study approval, IRB software by iMedRIS is an advanced, electronic research administration solution. Our IRB software is ideal for use by researchers and review boards alike at public and private research institutions.

Experience the most complete "out of the box" IRB software system on the market

- Use IRB software throughout the study approval process, from creating applications to study completion
- Electronically submit applications and related forms to specific review boards and committees
- Correspond electronically with researchers and committees
- Manage meeting times, agendas, minutes, and votes
- Merge comments, recommendations, and stipulations into electronic minutes
- Quickly create, submit, or approve applications, consent forms, and other study documents
- Streamline time-intensive routing and reviewing processes by routing forms through the system via email alerts
- Speed up the clinical trial reporting process with efficient, smoothly-running IRB software

IRB Manager (http://www.irbmanager.com/)

BEC is a software and services company focused on managing the information about the drug and medical device development process—from discovery through development and into the marketplace.

Our flagship product, IRBManager, is aimed at supporting the IRB Administrator and their staff as they struggle to control and organize more and more information in less and less time. Through concepts like Automated Event Management, One-Click Access and Investigator Self Service IRBManager helps lessen the "administrative" tasks with which an IRB Administrator has to deal. IRBManager is a browser based system available in both a Software as a Service (SaaS) and an on-site licensing model.

IRBNet (https://www.irbnet.org/release/index.html)

The Industry's Most Complete Solution

IRBNet's unmatched suite of electronic solutions drives compliance and productivity for your Administrators, Committee Members, Researchers and Sponsors. These powerful research design, management and oversight tools support your IRB, IACUC, IBC, COI and other Boards with a unified solution.

Flexible, Intuitive and Easy to Use

Your own forms. Your own processes. Your own standards. Powerful reporting and performance metrics. The data you need. From electronic submissions to form wizards, to agendas, minutes, and more. Our easy to use, web-based tools are rapidly launched and backed by our best practices expertise and the industry's leading support team.

Secure, Reliable and Cost-Effective

IRBNet's secure web-based solution is accessible to your research community anytime, anywhere. Our enterprise-class technology is cost-effective and designed to accommodate institutions of any size.

IRB+ (https://www.irbplus.com/)

IRB+ is the most affordable IRB database with online submission!

The new Online Submission Module is now available! PIs and Coordinators can

- Submit protocols applications, continuing reviews, amendments, adverse events, etc.
- Upload supporting documents like the consent form, full protocol description, advertising material, etc.
- View their activity letters online and respond online
- Check which IRB meeting their studies are scheduled for

You, as the IRB Administrator, can

- Create and change your own online forms
- Type less since the PIs and Coordinators are doing the data-entry for you
- Offer better service to your PIs through our 24×7 website

ProIRB (http://www.proirb.com/)

Our 2 Part Customized Software is the Only System That Grows with Your IRB!

Choose the System that Over 100 Institutions Use! Our 18 years experience working closely with IRB Administrators has allowed us to create software that anticipates the needs of the IRB! The ease of use of our products is why over 100 other Social Behavorial, Medical, Hospital and University clients have chosen us! Why Not YOU?!

Big or Small, We Have the Right-Sized System for You!

Our Customers range from IRB's with 3 protocols to 3000! With our unique 2 part system, allow your IRB staff to begin tracking your protocols with our affordable ProIRB® product. We grow with you! Add a New User or Additional Board at any time! Add Electronic Submission benefits for your PI's with CyberIRB® at any time! Implement the benefits of Electronic Submission with CyberIRB® logically and seamlessly with as few or as many PI's or Application Forms as you wish—NOT an all or nothing system (that could leave you Nothing!).

MOBILE TECHNOLOGY

Mosio (Mobile Solutions for Clinical Trials) (https://www.mosio.com/)

Mosio's mission is to help health and clinical research stakeholders maximize the power of mobile devices in an increasingly mobile world. We aim to make mobile technology simple for our clients and their end users, creating applications that increase patient engagement and improve outcomes through two-way communications, SMS alerts, surveys, and our own unique Storyline Alerts.

Our objectives are met through the ongoing development improvements of our platform by listening to customer feedback and testing new ideas. Our commitment to quality and customer satisfaction is enforced by an effective, thorough software development process and a formal quality management system based on HIPAA and 21 CFR Part 11 compliance requirements.

The Mosio team is committed to improving our performance in every aspect of our business, including privacy, data security, technological innovation, and ease of use in order to evolve healthcare and research on mobile devices.

Privacy and Data Security

Patient privacy and data security is our top priority. We realize the importance of providing a service enabling study participants to use their own devices, while maintaining levels of security and compliance requirements for HIPAA and 21 CFR Part 11 standards. Mosio is dedicated to ensuring private and secure software solutions with its clients, including Business Associate Agreements. If your organization requires a BAA, please email documents to hipaa((@))mosio.com and allow up to 7–10 business days for review and completion after the project/program agreements have been confirmed. Please contact us to receive our Privacy and Data Management Security statement.

PROFESSIONAL ORGANIZATIONS

Association of Clinical Research Professional (ACRP) (http://www.acrpnet.org)

ACRP supports clinical research professionals through membership, training and development, and certification. Founded in 1976, ACRP is a Washington, DC-based non-profit organization with more than 13,000 members who work in clinical research in more than 70 countries.

ACRP's vision is that clinical research is performed ethically, responsibly, and professionally everywhere in the world. ACRP's mission is to promote excellence in clinical research.

Vision

Clinical research is performed ethically, responsibly, and professionally everywhere in the world.

Mission

ACRP promotes excellence in clinical research.

Goals

(1) Serve as the preferred source for quality tools, resources, and best practices that support the clinical research community. (2) Serve as the gateway for the exchange of ideas and expertise across the clinical research community. (3) Champion the interests and perspectives of the clinical research community.

Association of Graduate Regulatory Educators (AGRE) (http://www.agreglobal.org/)

The Association of Graduate Regulatory Educators developed from a series of meetings, initiated in 2010, that brought together leaders of programs offering graduate training in regulatory affairs and regulatory science. From these meetings came the agreement to develop a formal organization with several goals:

AGRE is a place to meet and exchange ideas with other regulatory educators internationally.

AGRE is a forum for coordinating input on policy issues of importance to our educational programs.

AGRE is a group that develops and consolidates teaching materials to make your teaching more effective.

AGRE is a community for research on competencies, advancement of the discipline, and development of the profession.

Consortium of Academic Programs in Clinical Research (http://www.coapcr.org/)

The Consortium of Academic Programs in Clinical Research (CoAPCR) facilitates the development of high-quality educational programs encompassing all areas of clinical research that are based in academic credit-granting institutions.

CoAPCR's Mission

- To provide a medium for communication among educators of clinical research professionals.
- To encourage and support the development and maintenance of academically based clinical research educational programs to meet the needs of the clinical research community.
- To foster inter-institutional articulation among educational institutions, clinical institutions, professional associations, and industry.
- To initiate and/or support research and studies relating to the educational, manpower and service needs of clinical research professionals.

Core Competencies and Accreditation

A major aim of CoAPCR is to explore accreditation for academic clinical research educational programs. Discussions on this topic have taken place at several national meetings including ACRP, APCR, IACRN and DIA Annual meetings. More specifically, CoAPCR has dedicated a systematic approach to identifying the knowledge, skills and attitudes (KSAs), in the form of competencies that are necessary for clinical researchers to achieve successful professional outcomes.

CoAPCR Core Competency Domains

- Scientific Concepts and Principles of Research Design
- Medical Product Development
- Ethical Considerations and Responsible Conduct of Clinical Research
- Clinical Study Operations and Regulatory Compliance
- Study and Site Management
- Data Management and Informatics
- Communication of Scientific Data
- Professionalism, Teamwork and Leadership

Drug Information Association (DIA) (www.diaglobal.org)

For more than 50 years, DIA (the Drug Information Association) has served as a global forum for all those involved in health care product development and life cycle management to exchange knowledge and collaborate in a neutral setting. DIA is an essential resource that provides opportunities to extend debate and discussion to advance scientific and medical innovation.

DIA fosters innovation to improve health and well-being worldwide by:

- Providing invaluable forums to exchange vital information and discuss current issues related to health care products, technologies, and services;

- Delivering customized learning experiences;
- Building, maintaining, and facilitating trusted relationships with and among individuals and organizations that drive and share DIA values and mandates; and
- Offering a multidisciplinary neutral environment, respected globally for integrity and relevancy.

DIA Vision (http://www.diaglobal.org/)

DIA is your essential partner in catalyzing knowledge creation and sharing to accelerate health care product development.

DIA combines business principles, an understanding of the good clinical research practices, and skills necessary for navigating the changing political, regulatory, and financial landscapes.

The Certificate Program will provide the fundamental skills you need to do your job. Topics include, but are not limited to:

- Overview of drug development
- Clinical statistics for nonstatisticians
- Clinical project management
- Development of a clinical study report
- Art of writing a clinical overview
- Oversight of clinical monitoring: trends and strategies

International Conference on Harmonisation (ICH) (http://www.ich.org/home.html)

The International Council for Harmonisation of Technical Requirements for Pharmaceuticals for Human Use (ICH) is unique in bringing together the regulatory authorities and pharmaceutical industry to discuss scientific and technical aspects of drug registration. Since its inception in 1990, ICH has gradually evolved, to respond to the increasingly global face of drug development. ICH's mission is to achieve greater harmonisation worldwide to ensure that safe, effective, and high quality medicines are developed and registered in the most resource-efficient manner.

ICH's mission is to make recommendations towards achieving greater harmonisation in the interpretation and application of technical guidelines and requirements for pharmaceutical product registration, thereby reducing or obviating duplication of testing carried out during the research and development of new human medicines.

Launched in 1990, ICH is a unique undertaking that brings together the drug regulatory authorities and the pharmaceutical industry of Europe, Japan and the United States.

Regulatory harmonisation offers many direct benefits to both regulatory authorities and the pharmaceutical industry with beneficial impact for the protection of public health. Key benefits include: preventing duplication of clinical trials in humans and minimising the use of animal testing without compromising safety and effectiveness; streamlining the regulatory assessment process for new drug applications; and reducing the development times and resources for drug development.

Harmonisation is achieved through the development of ICH Tripartite Guidelines. The Guidelines are developed through a process of scientific consensus with regulatory and industry experts working side-by-side. Key to the success of this process is the commitment of the ICH regulators to implement the final Guidelines.

International Society for Biological and Environmental Repositories (ISBER) (http://www.isber.org/)

ISBER is the only global forum that addresses harmonization of scientific, technical, legal, and ethical issues relevant to repositories of biological and environmental specimens.

ISBER fosters collaboration; creates education and training opportunities; provides a forum for the dissemination of state-of-the-art policies, processes, and research findings; and provides an international showcase for innovative technologies, products, and services. Together, these activities promote best practices that cut across the broad range of repositories that ISBER serves.

Mission

ISBER is a global organization which creates opportunities for sharing ideas and innovations in biobanking and harmonizes approaches to evolving challenges for biological and environmental repositories.

Vision

ISBER will be the leading global forum for promoting harmonized high quality standards, ethical principles, and innovation in the science and management of biorepositories.

National Board of Medical Examiners (NBME) (http://www.nbme.org/NewInitiatives/clinicalresearchprogram.html)

Founded in 1915, the National Board of Medical Examiners (NBME) is an independent, not-for-profit organization that serves the public through its high-quality assessments of healthcare professionals.

The NBME research enterprise provides new information for use in enhancing existing products, developing new products, informing best practices, and aiding evidence-based decisions at both programmatic and institutional levels. While NBME's research is spread throughout the organization, prioritizing and managing research is the responsibility of the Office of Research (TOR) and the Research Implementation Committee (RIC). Together, TOR and RIC ensure the alignment of NBME research with institutional priorities. Current projects include investigations related to the assessment of new constructs, test score scaling and equating, score reporting and feedback, validity, group differences, and other areas.

Overview

The introduction of a Clinical Research Program aligns with the NBME's mission to protect the public by assessing the knowledge of individuals who conduct research in the field. Clinical research studies are carefully designed to determine the safety and effectiveness of better ways to prevent, detect or treat diseases.

NBME's Clinical Research Program (CRP) was established to aid personnel in the field of clinical research by providing a globally recognized, high-quality certification program. The examinations are designed to provide a baseline for a universally recognized, unbiased measure of the foundational knowledge of clinical research professionals to perform safe, effective research in human subjects. The CRP offers two certification examinations for responsibility-defined roles in the conduct of research:

- Monitor, Associate, and Coordinator Certification Examination
- Investigator and Scientist Certification Examination

Goals of the Program

The goal of the Clinical Research Program is to provide the following benefits to a spectrum of organizations within the clinical research field:

- Establishment of a baseline of required knowledge;
- Standardized documentation of core competencies and skills;
- Identification of competency gaps;
- Enhancement of training, education, and remediation; and
- Enhancement of professional development, recognition, and career mobility.

Public Responsibility in Medicine and Research (PRIM&R) (http://www.primr.org)

Since its founding in 1974, PRIM&R has pursued two core goals: creating a strong and vibrant community of ethics-minded research administration and oversight personnel, and providing educational and professional development opportunities that give that community the ongoing knowledge, support, and interaction it needs to raise the bar of research administration and oversight above regulatory compliance. PRIM&R has also formalized professional standards and credentials that document the expertise necessary for effective and appropriate research ethics support, and is active in public policy, offering expert opinion and guidance to the rule-making and advisory bodies governing the research enterprise and its practical applications.

PRIM&R's educational and professional development programs address a range of issues surrounding research involving human subjects and animals. Our approach welcomes the diversity of reasonable viewpoints, while ensuring the dissemination of thoughtful and accurate information. Much of our educational programming is taught by professionals with decades of experience in their respective fields.

Mission

Public Responsibility in Medicine and Research (PRIM&R) advances the highest ethical standards in the conduct of biomedical, behavioral, and social science research. We accomplish this mission through education, membership services, professional certification, public policy initiatives, and community building.

Vision

PRIM&R envisions a world in which all stakeholders in the research enterprise share an understanding of and commitment to the centrality of ethics to advancing science and medicine, as exemplified through research policies and practices that align with the highest ethical standards in research.

Core Values

We believe that "values" replace "rules" in the life of an organization; values exemplify what actions and behaviors the organization is trying to model, as well as those that it hopes its community will embrace. PRIM&R's core values are as follows:

1. Excellence—We strive to be capable and effective at what we do, and we maintain a commitment to excellence in all that we do.
2. Community—We recognize the connectedness and inter-reliance of all people; the value of teamwork and cooperation; and the importance of finding common ground in our interactions with one another.
3. Diversity—We value and promote the diversity of people, ideas, and opinions.
4. Integrity—We work to ensure that our practices are always aligned with our beliefs and that we are honest in our actions and in our words.
5. Knowledge—We seek understanding and intellectual stimulation through high-quality education.
6. Respect—We show consideration and courtesy towards all people and their perspectives.
7. Social Responsibility—We strive to be a good organizational citizen by maintaining awareness of social problems and global challenges and by doing our part to ameliorate them.
8. Creativity—We encourage the development of innovative ideas, strategies, and programming.

Society of Clinical Research Sites (SCRS) (www.myscrs.org)

The Society for Clinical Research Sites was founded in 2012 in response to the growing need for a trade organization to represent the voice of research sites within the clinical research enterprise. SCRS provides sites with resources, mentorship, and new ideas through a membership organization dedicated to providing the sites both a voice and community.

SCRS' Mission

Unify and amplify the voice of the global clinical research site community for site sustainability.

Advocate. SCRS will, on behalf of its site members, publicly comment on issues relevant to the clinical research industry. SCRS will work to maintain a presence in conversations with other groups and organizations to ensure the site's perspective and voice is present as important decisions and debates are occurring.

Connect. SCRS provides a unique opportunity to sites and industry stakeholders to interact with companies and individuals across multiple sectors of the industry. The SCRS community creates an environment that supports meaningful partnerships all year long and not just at meetings.

Educate. SCRS' knowledge platform creates the foundation for informed action, where industry leaders can analyze their most efficient programs, fresh voices can introduce proven innovations, and established experts can debate different interventions. SCRS and its members provide each other with a constant flow of new solutions and lessons learned.

Mentor. Recognizing both the importance of SCRS' emphasis on excellence and the reality that new sites enter the industry each day, site members will offer mentoring to other site members.

Society of Clinical Research Associates (SOCRA) (http://www.socra.org/)

Mission and Introduction

The Society of Clinical Research Associates (SOCRA) is a non-profit, charitable and educational membership organization committed to providing education, certification, and networking opportunities to all persons involved in clinical research activities. SOCRA, the premier educational organization for oncology site coordinators, has now emerged as a leading educational organization for clinical researchers in all therapeutic areas, supporting industry, government and academia.

Since incorporation in 1991, SOCRA has been through many changes, all of which were important contributors to our growth. The path that brought us to this level started with an exploration into available educational opportunities for site based clinical coordinators. The lack thereof and the thirst for information resulted in an organization founded by creative and forward thinking leaders. Today, the organization has realized membership growth and program expansion exceeding its expectations.

The most important factors in our success over the past years have been our membership support, our educational programming and our certification program. Innovation and investment of skill and knowledge have resulted in an exceptional organization with expertise and understanding in providing educational programming and member services. The quality of our programs and educators is unparalleled.

Regulatory Affairs Professionals Society (RAPS) (http://www.raps.org/)

The Regulatory Affairs Professionals Society (RAPS) is the largest global organization of and for those involved with the regulation of healthcare and related products, including medical devices, pharmaceuticals, biologics and nutritional products. Founded in 1976, RAPS helped establish the regulatory profession and continues to actively support the professional and lead the profession as a neutral, non-lobbying nonprofit organization. RAPS offers education and training, professional standards, publications, research, knowledge sharing, networking, career development opportunities and other valuable resources, including

Regulatory Affairs Certification (RAC), the only post-academic professional credential to recognize regulatory excellence. RAPS is headquartered in suburban Washington, DC, with offices in Shanghai and Singapore and chapters and affiliates worldwide.

Mission

Develop and sustain a competent global regulatory workforce that drives good regulatory practice and policy to advance public health.

Society for Clinical Data Management (SCDM) (http://scdm.org/)

SCDM is a non-profit, international organization created to advance the discipline of Clinical Data Management. SCDM members are charged with promoting quality and excellence in data management and are dedicated to the development, support and advancement of Clinical Data Management professionals.

SCDM strives to become the discipline's world leading advocate.

SCDM is organized exclusively for educational purposes. Through collaborations and partnerships, the Society creates a network of professionals driving the industry forward and create new offerings that will support Clinical Data Management professionals, from all horizons and career levels.

The Society also produces some key publications:

- The award-winning standard *Good Clinical Data Management Practices* (GCDMP©)
- The quarterly digital peer-reviewed journal *Data Basics* and
- The bi-monthly e-newsletter *Data Connections*

SCDM has created a certification program (CCDM™) to further support the profession.

TRAINING

Barnett International (http://www.barnettinternational.com/)

Superior clinical research training courses and training consulting services

- CORE CURRICULUM

  Barnett's conversion of our live seminars to a more flexible format, allows participants the flexibility to participate in the "core curriculum" offerings either in-person or via the web.
- INTERACTIVE WEB SEMINARS

  A Barnett Interactive Web Seminar offers you a seamless, secure, multimedia learning experience. These live, instructor-led sessions are designed to be highly interactive, and can be attended by individual attendees or groups at one low cost. No travel, no travel expenses, and no time away from the office!

- WEB SEMINAR ARCHIVES
  Were you unable to attend an Interactive Web Seminar? These DVD archives will allow you to watch recordings of previous Interactive Web Seminars any time you want. Pricing is available for single users and site licenses.
- PUBLICATIONS
  Our reference manuals help research facilities ensure compliance by providing updates on the latest Federal regulations, while our industry compendiums provide executives with valuable information garnered from real-world studies, analyses, and fresh insight from widely respected opinion leaders on the most important new developments in the industry.
- ON-SITE TRAINING
  Available worldwide, at your location, Barnett International's On-Site Seminars are a more cost-effective way to train a group of employees than other training alternatives.
- TRAINING STRATEGY & CONSULTING
  How do you ensure that your training programs can adequately withstand FDA scrutiny? Reviews of FDA 483s and warning letters indicate that the most frequently issued process deficiencies include areas that can be easily addressed with focused training programs.
- GCP TRAINING & ASSESSMENT
  Using a rigorous test question development and validation process, Barnett provides formal Good Clinical Practice (GCP) training and assessment for global clinical research professionals. Find out why companies are viewing Barnett's assessment as a new standard for GCP Certification.
- eLEARNING
  Does your department have critical training needs that need constant reinforcement? Barnett's customized eLearning development services allow you to train large groups of employees in a consistent and cost-effective manner. Designed as self-paced modules, Barnett's eLearning programs offer highly interactive, fun and engaging learning experiences for your teams.

Collaborative Institutional Training Initiative (CITI) (https://www.citiprogram.org/)

> To promote the public's trust in the research enterprise by providing high quality, peer reviewed, web based, research education materials to enhance the integrity and professionalism of investigators and staff conducting research.
>
> *CITI Program Mission Statement*

Paul Braunschweiger Ph.D., Professor, Radiation Oncology at the University of Miami Miller School of Medicine, and Karen Hansen Director, Institutional Review Office at Fred Hutchinson Cancer Research Center, founded the Collaborative Institutional Training Initiative (CITI Program) in March 2000. Representatives from the Albany Medical Center, the Children's Hospital of Boston, Dartmouth College, the Fred Hutchinson Cancer Research Center, the Group Health Cooperative, the National

Comprehensive Cancer Network, the University of Kentucky, the University of Miami, the University of Nebraska Medical Center, and the University of Washington comprised the first group of content experts charged with developing the first CITI Program course focused on human subjects research (HSR) protections.

The CITI Program's biomedical HSR content was expanded significantly in 2004 to include content for social and behavioral researchers (SBR). The initial SBR development group were representatives from the American Psychological Association, Columbia University, Duke University, Mississippi State University, the North Carolina Department of Corrections, the Research Triangle Institute, Rutgers University, the University of Chicago, and the University of Virginia.

CITI Program Advisory Committee, led by Jaime Arango Ed.D, CIP, includes many teams of expert writers, editors, and reviewers, who create and manage compliance training materials in the areas of: *Animal Care and Use* (ACU), *Biosafety and Biosecurity* (BSS), *Clinical Research Coordinator* (CRC), *Clinical Trial Billing Compliance* (CTBC), *Conflicts of Interest* (COI), *Disaster Planning for the Research Enterprise* (DPRE), *Essentials of Research Administration*, *Export Compliance* (EC), *Good Clinical Practice* (GCP), *Good Laboratory Practice* (GLP), *Healthcare Ethics Committee* (HEC), *Human Subjects Research* (HSR), *Information Privacy and Security* (IPS), *IRB Administration*, and *Responsible Conduct of Research* (RCR). These materials reach more than a million leaners annually at thousands of research institutions.

The CITI Program joined the Biomedical Research Alliance of New York (BRANY) in May 2016 in order to better address the educational needs of investigators, staff and students, in the global research community.

HealthCarePoint (https://www.healthcarepoint.com/)

HealthCarePoint's BlueCloud® is a universal networking technology that enables industry stakeholders to innovate and create new and efficient business models that improve patient care, by implementing standards and eliminating waste, fraud and abuse. Its network reaches more than 162 countries and is used by nearly 1,000,000 healthcare professionals, 45 Sponsors, 14 CROs, 8 IRBs, dozens of universities and thousands of healthcare and clinical research stakeholders and site organizations. Dozens of BlueCloud® applications work together as one, empowering all industry stakeholders to organize, centralize, connect, share and deliver information required for business and compliance by using a common, secure and private verifiable on demand system, saving the industry time, money and ultimately saving lives.

Mission

To connect healthcare and clinical research by empowering industry stakeholders to link, share and exchange information in "real-time" in order to save time, save money and ultimately save lives.

Appendix D: Apply Your Knowledge

CHAPTER 1—RULES, ROLES, AND RESPONSIBILITIES

1. PharmaXYZ wants to place a diabetes study at a research site that specializes in dermatology. The key question to be asked is whether the investigators at the dermatology site have the qualifications, training, education, and expertise to conduct a diabetes study. Most likely, because they are in a dermatology practice, they do not have the expertise to conduct a diabetes study. The sponsor is responsible for selecting qualified investigators, the investigators are responsible for reading and understanding the study information as well as conducting the study, and the institutional review board (IRB) is responsible for ensuring the protection of the rights and welfare of human subjects.
2. Investigator ABC has a grant from the Federal government to conduct a study that also has an Investigational New Drug (IND). This study falls under the Common Rule (45 C.F.R. 46) because the study is funded by the Federal government and Food and Drug Administration (FDA) regulations (21 C.F.R. 50) because the sponsor of the study (who may be working with the National Cancer Institute [NCI]) has applied for an IND. Investigator ABC and her research site should follow both sets of regulations.

CHAPTER 2—PRODUCTS, PROTOCOLS, AND PRETRIAL PREPARATION

1. Dr. Trail is reviewing a study for feasibility when she discovers that the 30-day IND review by the FDA is still ongoing. This is of concern because the protocol is "technically" not in its final form. The FDA may request the sponsor to make numerous changes and/or may not approve the study in its current form. As such, the study may change significantly and Dr. Trail may need to start the feasibility process again.
2. Dr. Marsh has two other studies that are almost identical to the one he is currently reviewing. In his review, Dr. Marsh most likely will focus on whether the new study has scientific merit. Because he is already conducting 2 studies that are almost identical, what added knowledge from the new study will be relevant to his patients or to generalized knowledge? Also, Dr. Marsh will most likely consider whether it is ethical to offer his patients another study that is so similar to his current offerings, and he will consider whether he has enough resources to offer another study that is so similar to his current offerings.

3. During a feasibility review of a protocol, Dr. Callen discovers that preclinical trials of the investigational product indicated severe toxicities in various animal studies. Although animal species differ in reactions to various substances, Dr. Callen has a legitimate concern in regard to risk. He should contact the sponsor to gain insight on his concerns and to gain additional background on the study. There may be different reasons why the protocol does not include the information, including an omission during the drafting of the protocol and informed consent documents. Keeping in mind that Dr. Callen is responsible for obtaining IRB approval of the study and the ultimate conduct of the study at the research site, he must be able to communicate the risks and benefits of a study to potential subjects.

CHAPTER 3—SPONSOR, SITE, AND STUDY START-UP

1. The investigator is too busy and has asked his or her study coordinator to be responsible for the oversight of a study. Although the study coordinator may be capable of conducting the study, the investigator is legally (through Form FDA 1572 and the clinical trial agreement [CTA]) responsible for the conduct and oversight of the study. The investigator may assign duties to qualified study staff via the delegation of authority; however, she or he may not delegate oversight of the trial. The coordinator should graciously decline and explain the legalities while letting the investigator know that he or she will support him or her in any way possible. The coordinator is not able to make medical decisions, determine adverse event criteria, or provide medical treatment to the subject. Another option would be for the investigator to not agree to conduct the study as she or he does not have the time needed to commit to the trial.
2. The site is well on its way to being ready for subject screening. However, there are a few items missing that need to be addressed before the study can start. These include the following:
   a. Installment and training of the electronic data capture system.
   b. Having signatures on file for electronic signatures.
   c. Completed conflict of interest forms for study staff.
   d. Confidentiality agreements signed by study staff.
   e. The budget must be approved.
   f. The CTA must be approved.
   g. A recruitment plan should be created.
   h. Any recruitment or advertising materials must be approved by the IRB.

CHAPTER 4—ENTICEMENT, ENROLLMENT, AND ENGAGEMENT: THE INFORMED CONSENT PROCESS

1. Mr. Marsh stops taking his medication without asking his doctor, in anticipation of joining a clinical trial. This may be dangerous for Mr. Marsh because he may need the medication to function, or he may suffer withdrawal symptoms or other side effects. Although he thinks that he qualifies

for the study, Mr. Marsh may not understand that the study may have other specific enrollment qualifications and/or that the study may be closed to enrollment. Investigators and research site staff have little impact on decisions of individuals like Mr. Marsh but should be aware that this occurs and advise individuals who are interested in studies to wait until it is determined that they qualify for the study before taking any action and to take action only under the supervision of a health care provider.

2. Dr. Andrews wishes to review a clinic's database to determine if her study outline will be a good fit for an asthma patient population. Most likely, her request will fall under "Activities preparatory to research" and she needs to agree with the covered entity that the use is solely in preparation for research, that the protected health information (PHI) will not be removed from the covered entity in the course of review, and that the PHI is necessary for the research.

CHAPTER 5—FROM ENROLLMENT TO FINAL VISIT

1. The informed consent form was not signed. This is considered a violation and must be documented in the subject binder and reported to the sponsor and the IRB as soon as possible. You would need to call the subject and explain the situation. Ask the subject to come back to the site as soon as possible and have him or her sign the consent. You could have the subject sign on the current date with a note that they gave verbal consent on the date of screening. A note to file should document that consent was given at screening, subject questions were answered, and the subject agreed to participate in the study. During the consenting process, the research assistant forgot to obtain a signature from the subject. The subject was called immediately upon notice of this violation and came in on __/__/____ to officially sign the informed consent form. As a result of this incident, all staff were retrained on the informed consent process and reviewed and signed the site's informed consent process standard operating procedure.
2. To comply with 21 C.F.R. 11, the software must be secure and track all transactions that occur. It must include a double-secure sign-on for each user, which includes the user's name and password. The software should allow access to the various components based on the role of the user. For example, a coordinator may have more access rights than a lab tech does. The system must track any change made to the data and identify the person who made the change, the date and time of the change, and the reason for the change. Food and Drug Administration controls and requirements include the following:
   a. Limiting system access to authorized individuals
   b. Use of operational system checks
   c. Use of authority checks
   d. Use of device checks
   e. Audit trail of all user activities
   f. Determination that persons who develop, maintain, or use electronic systems have the education, training, and experience to perform their assigned tasks

CHAPTER 6—COLLABORATING FOR COMPLIANCE AND QUALITY DATA—MONITORING AND AUDITS

1. Subject 16 did not receive a computed tomography (CT) scan after Visit 2, required by the protocol. Using the 5 Whys, one possible root cause follows:
   a. Why did Subject 16 not receive a CT scan after Visit 2? He missed his CT appointment.
   b. Why did Subject 16 miss his CT appointment? He had no transportation to the appointment.
   c. Why did Subject 16 not have transportation to the appointment? His car was repossessed by the bank for missing multiple payments.
   d. Why was his car repossessed? He used his car payment money for medication.
   e. Why did he use his car payments for medication? He needs to take expensive medication and he does not have the budget for both medication and car payments.
2. A corrective action/preventive action (CAPA) plan for Subject 16's failure to obtain a CT exam might include ensuring that he has transportation to and from appointments through car services or other transportation. Although the root cause of this situation is that Subject 16 cannot afford his car, the solution for this root cause is outside the scope of the research site's influence. In this case, the research site staff can suggest a CAPA that includes finding alternative transportation for subjects.

CHAPTER 7—BUILDING BUDGETS

1. No. There are several items missing that need to be included in the budget. Some of these may be included in the invoicable items.
   a. Study site start-up fee
   b. Cost for investigator meeting for investigator and coordinator
   c. Cost for screen failures
   d. Special equipment
   e. Monitor visit costs
   f. Inspection/audit fees
   g. Labs and processing of labs
   h. Cost to adjudicate serious adverse event
2. Justify start-up costs by providing historical site information on the activities, staff, and time that it takes the site to properly set up and prepare the site for the study. Provide metrics showing that the site has been prepared and ready to go for past studies and has conducted studies efficiently and effectively, meeting deadlines and study goals. The sponsor compensates their representatives for their time to evaluate and initiate the site. The site should also be compensated fairly for their time. Because the site would not have this cost if they were not conducting the study, it is not a cost of doing business but a research-specific cost.

If the sponsor expects the coordinator to be available during a monitoring visit, it should be willing to compensate the coordinator's time. In order to be available at specific times or during the monitor's visit, the coordinator will have to arrange time away from other tasks or studies that she or he is responsible for. Again, this is a specific cost of doing research and should be compensated as such. With careful planning and management, the sponsor can limit the time required and the expense will not be substantial.

CHAPTER 8—CONTRACTS, CLAUSES, AND CLOSING THE DEAL

1. Replace the word *immediately* with *in a timely manner.* In addition, add details specifying who the report should be sent to and how it should be sent (fax, email, etc.)
2. The CTA should state that the contract may be terminated by either party. It should provide reasons for termination, define how much notice is required prior to termination, and how the notice must be communicated (certified mail, fax, via phone, email) and who is to be notified of the termination.

CHAPTER 9—U.S. CLINICAL TRIALS—ADDITIONAL TOPICS

1. The two categories are as follows:
   a. Stem cells that are "introduced into non-human primate blastocysts" or
   b. Research that involves the "breeding of animals where the introduction of hESCs…or human induced pluripotent stem cells may contribute to the germ line."

   The National Institutes of Health is hesitant to include these categories due to the ethical and scientific issues and there is not public consensus in this area.
2. The primary mode of action (PMOA) determines the lead center, whether a drug or device.
   a. PMOA—Center for Drug Evaluation and Research (CDER)/medication
   b. PMOA—Center for Devices and Radiological Health/stent opening the artery
   c. PMOA—CDER/medication

Index

Printed in the United States
by Baker & Taylor Publisher Services